Microorganisms in Foods, Volume 2

Sampling for Microbiological Analysis: Principles and Specific Applications

Second Edition

International Commission on Microbiological Specifications for Foods

This thoroughly revised, updated, and much expanded edition of *Microorganisms in Foods 2* retains the basic format and purpose of the first edition published in 1974. It preserves the statistical treatment, but Part I (Principles) now includes a discussion of the Hazard Analysis Critical Control Point concept and Part II (Specific Proposals for Sampling and Sampling Plans) has been reorganized into sixteen commodity categories: raw meats; processed meats; poultry and poultry products; feeds of animal origin and pet foods; milk and milk products; eggs and egg products; fish and shellfish; vegetables, fruits, and nuts; soft drinks, fruit juices, concentrates, and fruit preserves; cereals and cereal products; spices, condiments, and gums; fats and oils; sugar, cocoa, chocolate, and confectionery; formulated foods; natural mineral waters, other bottled waters, process waters, and ice; and shelf-stable canned foods. In addition, Part I has been revised to include microbiological hazards which have only recently come to light.

This new edition will be of even greater value than the first to food microbiologists and those responsible for food quality control and public health.

The International Commission on Microbiological Specifications for Foods, a standing commission of the International Union of Microbiological Societies, was formed in 1962 in response to the need for international microbiological specifications for foods. Its members are internationally known food microbiologists, drawn from government laboratories in public health, agriculture, and food technology, and from universities and the food industry.

SPONSORED BY

The International Commission on
Microbiological Specifications for Foods,
of the International Union
of Microbiological Societies

EDITORIAL COMMITTEE

T.A. Roberts (Chairman), F.L. Bryan, J.H.B. Christian,
D. Kilsby, J.C. Olson Jr, and J.H. Silliker

CONTRIBUTORS

ICMSF *Members*

J.H.B. Christian (Chairman),
F.L. Bryan (Secretary), J.H. Silliker (Treasurer),
A.C. Baird-Parker, D.S. Clark, O. Emberger,
Betty C. Hobbs, Milica Kalember-Radosavljecic,
J. Liston, P.M. Nottingham, F. Quevedo,
T.A. Roberts, B. Simonsen, H.-J. Sinell,
P. Teufel, R.B. Tompkin, M. van Schothorst

Former Members

K.H. Lewis, J.C. Olson, Jr,
C.F. Niven, Jr, H. Pivnick

Consultants

G. Jarvis, D. Kilsby

8/93

MICRO-ORGANISMS IN FOODS

2

Sampling for microbiological analysis: Principles and specific applications

Second edition

The International Commission on

Microbiological Specifications for Foods (ICMSF)

of the International Union of

Microbiological Societies

UNIVERSITY OF TORONTO PRESS

Toronto Buffalo London

First edition

Second edition

Toronto Buffalo London
Printed in Canada

ISBN 0-8020-5693-8

Printed on acid-free paper

Canadian Cataloguing in Publication Data

Main entry under title:

Microorganisms in food

Includes bibliographical references and index.
ISBN 0-8020-5693-8

1. Food – Microbiology – Technique. 2. Food – Analysis.
3. Sampling (Statistics). I. International Commission on
Microbiological Specifications for Foods.

QR115.M52 1986 576′.163′028 C85-099972-3

Contents

Tables

Figures

Preface to the second edition

In the first edition of this book (ICMSF, 1974; reprinted 1978), a group of food microbiologists and statisticians discussed the statistical aspects of sampling foods for microbiological analysis, presented the concept of 2- and 3-class attribute sampling plans, and suggested microbiological criteria for a range of food products and ingredients that move in international trade.

In subsequent years, these concepts have been increasingly embraced by the food industries and by national and international food-control authorities. In related activities, ICMSF undertook a series of international collaborative and comparative trials on media and methods for the isolation and enumeration of microorganisms in foods (Appendix 4) and at the request of the World Health Organization an ICMSF committee prepared a document on the application to food processing of the Hazard Analysis and Critical Control Point (HACCP) concept (WHO/ICMSF, 1982). Also during this period the ICMSF's first volume on laboratory methods was revised (ICMSF, 1978) and its third book (ICMSF, 1980a, b) was published. The latter book presents the principles of food microbiology and discusses in detail microbiological ecology in relation to both food processes and individual foods. It includes treatment of microbial hazards in foods and the reasoning behind the selection of cases for attributes sampling plans.

The purposes and organization of this book are described in detail in the preface to the first edition (see pp. xix–xxii below). The present edition preserves the statistical treatment, but Part II has been reorganized to follow the commodity categories in the ecology book (ICMSF, 1980b). Consideration of the 'General Principles for the Establishment and Application of Microbiological Criteria for Foods' (Codex, 1981), accumulated microbiological data, and experience in the use of the criteria cited in the first edition have resulted in either addition or deletion of criteria for some foods and modification of criteria for others.

These criteria are generally recommended for port-of-entry sampling and examination of foods. They are not necessarily appropriate to control of processing. However, guidelines are also suggested in some chapters to assist in process control in food plants.

The sampling plans suggested in this book are statistically based and derived from the best available microbiological data. In applying them, however, the microbiologist must make the final judgment based on experience and on the prevailing circumstances.

The Commission hopes that this edition will be of even greater value to food microbiologists than the first has been, and that they will continue to provide members of the Commission with advice and data on the microbiology of foods.

ACKNOWLEDGMENTS

The Commission is most grateful for advice and assistance from many scientists in the revision of this book, and in particular for the help of Mr D. Kilsby and Dr. G. Jarvis in statistical matters. Thanks are also due to the Editorial Committee chaired by Dr T.A. Roberts and to the chairmen of the chapter subcommittees.

This revision would have been impossible without financial sponsorship from many quarters, and the Commission is indebted to the World Health Organization and to many companies within the food industry for their generous support (see Appendix 3). Much of the revision was centred on workshops held in Stresa, Italy (1980), Chexbres, Switzerland (1981), and Anaheim, USA (1982) and on editorial discussions in Los Angeles, USA (1982), Montreux, Switzerland (1983), and Bedford, England (1983). Sincere thanks are extended to the hosts and organizers of these meetings.

Finally the Commission expresses its gratitude to various national governments, universities, and private companies for supporting the participation of their staff in the activities of the ICMSF.

REFERENCES

Codex 1981. *Codex Alimentarius Commission, Fourteenth Session, 1981: Report of 17th Session of the Codex Committee on Food Hygiene, Washington, D.C. 17–21 November 1980*. ALINORM 81/13. Appendix II, General Principles for the Establishment and Application of Microbiological Criteria for Foods. Codex Alimentarius Commission, FAO, Rome

ICMSF. 1974. *Microorganisms in Foods. 2. Sampling for Microbiological Analysis: Principles and Specific Applications.* 1st ed. University of Toronto Press, Toronto

– 1978. *Microorganisms in Foods. 1. Their Significance and Methods of Enumeration*, 2nd ed. University of Toronto Press, Toronto

– 1980a. *Microbial Ecology of Foods.* Vol. 1. *Factors Affecting Life and Death of Microorganisms.* Academic Press, New York

– 1980b. *Microbial Ecology of Foods.* Vol. 2. *Food Commodities.* Academic Press, New York

WHO/ICMSF. 1982. *Report of the WHO/ICMSF Meeting on Hazard Analysis: Critical Control Point System in Food Hygiene* (Document VPH/82. 37) WHO, Geneva, 9–10 June 1980

Preface to the first edition

Microbiological methods, no matter how accurate and reproducible, are inadequate to appraise the microbiological quality of food without a satisfactory sampling plan. Sampling plans in current use are often badly understood and highly unsatisfactory, and the level of safety they provide is generally not well known. At present, no comprehensive publication exists dealing with statistically based sampling plans, broadly applicable to the field of food microbiology. The facts that the production of foodstuffs is gradually being centralized in enlarging enterprises and that foodstuffs are moving in international commerce on a rapidly increasing scale make the need for such plans still more obvious. The Commission decided to try to fill this gap and, to that end, called upon expert statisticians and commodity specialists to supplement its own areas of competence.

The primary aim of this book is instructional: to draw attention to the unsatisfactory nature of casual sampling, and to provide information necessary to improve the situation. The immediate objectives are: to describe what different sampling plans can achieve in principle; to establish their significance for the interpretation of the results of microbiological testing of foods; and to offer guidance as to the choice of appropriate sampling plans in different situations.

The cost of sampling and testing and inadequate awareness of the principles involved nearly always preclude the use of an ideal sampling plan, and lead to the use of sampling procedures that are far from satisfactory in both nature and scope. Such a situation seems likely to produce an unjustified feeling of security in those responsible for the interpretation of the results. Improved sampling plans as indicated in this book will give a basis for statistically valid and realistic judgment of the microbiological quality of foods under test.

The Commission has limited its scope both in the previously published book and in the present text to the range of competence of its members and advisers. Thus neither includes the fields of parasitology, virology, and mycology. Incomplete survey data imposed additional limitations. The criteria proposed herein were based primarily on the knowledge of experts in the various commodity fields. They will apply to many situations, but must be tempered with good judgment in particular instances. The Commission intends to collect further information on these matters, which will probably make imperative a revision of this text in the relatively near future.

The scope and limitations, as outlined above, indicate that the book is intended primarily for people dealing with the microbiological safety of food moving in international and domestic commerce. National food control and public health authorities, food manufacturers, international organizations, and teachers in food microbiology should also find it useful.

The substance of the book falls into two parts: the principles of sampling (Part I), and the application of these principles to suggest sampling plans for various foods (Part II). Part I defines the statistical concepts and terms used in the book, describes the kind of sampling plans available and their purposes, outlines the procedures for selecting the best plan for the food and organism in question, and explains the principles of randomization. Part II describes practical applications of the principles set out in Part I and treats the procedures for collecting and handling sampling units for various food categories.

The text ends with general conclusions and a glossary of special terms.

The text is the result of five years of study by ICMSF members and consultants, involving workshops held in Dubrovnik, Yugoslavia (1969), Mexico City, Mexico (1970), Opatija, Yugoslavia (1971), Langford, England (1972), and Ottawa, Canada (1973). The project was proposed and planned by the chairman, F.S. Thatcher, after initial phases of collaboration with the chairman of the Statistics Committee, D.F. Bray. Major segments of the work were first studied by subcommittees and then discussed in plenary sessions. Results were compiled as minutes by the secretary, D.S. Clark, and later drafted into book form by the chairman, F.S. Thatcher. Subsequent compilation and rewriting was done by the various subcommittees. Editing was the responsibility of an editorial committee under Chairman M. Ingram.

The group represented a range of opinions and principles, whose expression required constructive compromise. For example, the impracticability at present of insisting that *Salmonella* be undetectable in some raw meats had to be reconciled with the reluctance to sanction the presence of a pathogen in any food. The sampling plans described herein are not inflexible standards. If applied with increasing stringency they should permit improvement in microbiological quality without making necessary the condemnation, in prohibitive proportion, of important foodstuffs.

NOTICE

The main aims of this book are to show what sampling plans can do and what their advantages are over casual sampling procedures. The examples set out in Part II are only intended to illustrate the organisms important in the different foods and how, for each, limits might be set. The Commission has not decided whether there is a sufficient need to apply all these limits. Moreover, the acutal limits chosen are based on information readily available to the Commission, which is mainly derived from a restricted range of countries and conditions; hence the suggested values are likely to need revision if applied to different circumstances. The values indicated are certainly not intended as hard and fast limits for universal application.

ACKNOWLEDGMENTS

Sincere thanks are extended to the many scientists who contributed to this book, in particular: Dr D.F. Bray, Professor I.W. Burr, and Dr E.F. Drion, who guided the Commission on statistical questions; Mr G.G. Anderson, Dr J. Liston, Dr C.F. Niven, and Dr J.M. Shewan, whose advice and assistance were much appreciated; the chairmen of the various subcommittees; and also the staff of the National Science Library of Canada, for confirming the accuracy of the reference citations.

The Commission is most grateful for generous financial sponsorship without which the work would have been impossible: from the United States Department of Health, Education and Welfare, Public Health Service, Health Services and Mental Health Administration, Center for Disease Control; the World Health Organization; Canada Department of National Health and Welfare, Health Protection Branch; and various companies within the food industry. This assistance does not, of course, constitute endorsement of the findings and views expressed herein. Finally, thanks are also expressed to the respective national governments, universities, and private companies for supporting the participation of their staff in the work of the Commission, of which the present text is but one result.

Abbreviations of societies and agencies

AMI	American Meat Institute
AOAC	Association of Official Analytical Chemists
APHA	American Public Health Association
CEC	Commission on the European Communities
DHSS	Department of Health and Social Security (UK)
EEC	European Economic Community (now European Communities)
FAO	Food and Agriculture Organization of the United Nations
FDA	United States Food and Drug Administration
FPI	Food Processors Institute
HPB	Health and Protection Branch, Health and Welfare Canada
HWC	Health and Welfare Canada
IAMS	International Association of Microbiological Societies
ICMSF	International Commission on Microbiological Specifications for Foods
IDF	International Dairy Federation
IEC	International Egg Commission
IOCC	International Office of Cocoa and Chocolate
ISO	International Standards Organization
IUBS	International Union of Biological Societies
IUMS	International Union of Microbiological Societies
NAS	National Academy of Sciences of the United States
NCA	National Canners Association
NCDC	United States National Communicable Disease Center
NCI	National Cheese Institute
NFPA	National Food Processors Association
NRC	National Research Council of the United States

NSDA	National Soft Drink Association
OSU	Oregon State University
USDHEW	United States Department of Health, Education and Welfare
USPHS	United States Public Health Service
WHO	World Health Organization of the United Nations

PART I
PRINCIPLES

1

Meaningful microbiological criteria for foods

Traditionally, three principal means have been used by government agencies and commercial organizations to control microbiological hazards of foods. These are: (a) education and training, (b) inspection of facilities and operations, and (c) microbiological testing. As in the previous edition of this book (ICMSF, 1978b), our primary emphasis is on use of microbiological testing through the application of microbiological criteria for foods – that is, standards, specifications, and guidelines – to minimize microbiological hazards of foods. However, we recognize that use of each of the above three traditional approaches is necessary to provide optimal control of microbiological hazards. Chapter 7 deals at length with a system that includes all of these approaches, the Hazard Analysis Critical Control Point (HACCP) system.

Since publication of the first edition of this book, interest in microbiological criteria has increased rapidly. This has stemmed primarily from the belief by many that microbiological criteria are useful as a means of ensuring the safety and quality of foods and thereby their acceptability to the consumer. Also, interest has been stimulated by the concern of food processors over the impact of criteria that may be imposed on them.

The rationale for application of microbiological criteria is that microorganisms bear a relationship to the safety and quality of food. The overriding factor, however, is safety. Safety is related to the presence or absence of pathogenic microorganisms or their toxins, the number of pathogens present, and the ease with which these agents can be controlled or destroyed. Microbiological quality is related to presence or absence of microorganisms that are capable of causing spoilage or are indicative of lack of Good Commercial Practice (GCP), and to the effectiveness of measures to control or destroy such organisms.

Application of microbiological criteria should distinguish between an acceptable and unacceptable product or between acceptable and unacceptable food processing and handling practices. The use of microbiological criteria for these purposes has been addressed previously by ICMSF and various individuals and groups, including several study groups of the World Health Organization (WHO) and the Food and Agriculture Organization (FAO) of the United Nations (ICMSF, 1978b, 1980; FAO/WHO, 1974, 1975, 1977, 1979; WHO, 1974, 1976; Codex, 1981a, b; NAS-NRC, 1964; Corlett, 1974; Foster, 1969; Kronick, 1973; Olson, 1975, 1978; Peterson, 1970; Thatcher, 1955). An attempt has been made here to bring into focus progress that has been made in this area of food microbiology and help place in perspective the content of subsequent chapters.

A. DEVELOPMENT OF PRINCIPLES FOR ESTABLISHMENT OF MICROBIOLOGICAL CRITERIA

Perhaps the most significant recent contribution towards development of meaningful microbiological criteria in foods was the preparation of the Codex Alimentarius 'General Principles for the Establishment and Application of Microbiological Criteria for Foods' (Codex, 1981a). Preparation of this document was prompted primarily by a resolution* from Technical Committee 34, Subcommittee 9, of the International Standards Organization (ISO) which expressed concern over the growing proliferation of needless and ill-advised criteria. It was originally drafted by an FAO/WHO Expert Consultation appointed at the request of the Codex Committee on Food Hygiene. Subsequently, the document was adopted with minor modification by the Codex Alimentarius Commission. *It is reproduced in full in Appendix 6*. Briefly, it sets forth: (a) the components of a microbiological criterion; (b) definitions for each of three types of microbiological criteria (*standard*, *specifi-*

* Recommendation of ISO, Technical Committee 34, Subcommittee 9, third annual session, Budapest, May 1976: 'Microbiological specifications for foods are being established by regulatory agencies throughout the world. There is concern among the knowledgeable that in many instances microbiological specifications are not based upon sound principles, for example, specifications have been set for certain foods for which no substantive need for such specifications can reasonably be justified; in other instances existing specifications reflect inadequate producer or consumer risks or both. Subcommittee 9 of ISO/TC 34 recommends that FAO/WHO act to convene a joint FAO/WHO Expert Committee to set forth guiding principles for the establishment and application of microbiological specifications for foods.'

cation, *guideline*); (c) the purposes of microbiological criteria; (d) the factors that must be considered to fulfil the purposes of microbiological criteria; (e) the basis for selection of each component of a microbiological criterion; and (f) actions to be taken based on results of application of criteria.

Although the 'General Principles' were intended to provide guidance for Codex purposes, national, regional, and local regulatory agencies and industry can easily adapt them to their own particular needs. ICMSF has done that in developing its recommendations of criteria for the food commodity groups given in Part II of this book. It is emphasized, however, that the criteria recommended were predicated on three prevailing premises: (a) the food should be in international trade; (b) the criterion is needed either for safety or for good commercial practice; and (c) the criterion would serve a useful purpose.

B. DEFINITIONS

Microbiological criteria were defined in the previous edition of this book, but more comprehensive definitions are now needed. ICMSF, while endorsing the definitions given in the 'General Principles' for Codex purposes, believes that some modifications are necessary to serve ICMSF needs. The definitions that follow are intended to reflect this viewpoint.

A *microbiological criterion* for food as defined for purposes of the ICMSF and as used herein consists of:

1 a statement of the food to which the criterion applies;
2 a statement of the contaminant(s) of concern (i.e., the species or groups of microorganisms and/or their toxins);
3 the analytical method(s) to be used for its (their) detection and/or enumeration;
4 the sample to be taken from a lot of food or from a source of concern such as a point in a processing line;
5 the microbiological limit(s) appropriate to the product; that is, the values of n, c, m, and M (see Chapter 5).

The term *sampling plan* as used in this book includes all of the above.

The terms to describe microbiological criteria for purposes of the ICMSF and as used in this book are defined as follows:

1 A *microbiological standard* is a microbiological criterion, as defined above, which is part of a law or regulation. It is a *mandatory*

criterion which is enforceable by the regulatory agency having jurisdiction.

2 A *microbiological specification* is a microbiological criterion, as defined above, which is applied as a condition of acceptance of a food or ingredient by a food manufacturer or a public or private agency. It is a *mandatory* criterion in the sense that foods or ingredients that do not meet the specification(s) are subject to rejection by the buyer. In the previous edition this was referred to as a purchasing specification.

3 A *microbiological guideline* is a microbiological criterion, as defined above, which is used by a manufacturer or regulatory agency to monitor a food process or system. Guidelines aid in determining whether or not microbiological conditions at critical control points or for the finished product are acceptable. Guidelines are usually *advisory* but may be *mandatory* depending upon the source of sample (e.g., critical control points), types of organisms of concern, and the nature of the food being produced or processed.

An explanation of some differences between these definitions and those in the 'General Principles' document (Codex, 1981a, reproduced in part in Appendix 6, p. 269) may be helpful. The Codex microbiological criteria are tied closely to Codex codes of hygienic practice. The term 'specification' is applied exclusively to an end product and is intended to increase assurance that provisions of hygienic significance in the respective code have been met. However, food processors, private and public institutions, and various government agencies, including the military, often require vendors or suppliers to provide foods and ingredients that do not exceed certain microbial limits as specified in purchase specifications. No mention of a microbiological criterion for such a purpose is made in the Codex 'General Principles' document. Microbiological guidelines are intended to monitor critical control points in an establishment or finished product. For Codex purposes they are intended only to guide the manufacturer, and specifications and guidelines are advisory criteria. Microbiological standards are intended to be Codex Alimentarius Standards (as distinct from being provisions in Codex codes of hygienic practice) and are mandatory. They shall not be introduced de novo but shall be derived from microbiological end-product specifications which have accompanied codes of practice through the Codex procedures and which have been extensively applied to foods. Presumably, the experience gained would justify either establishment of a standard or continuance of the criterion as an end-product specification.

C. SELECTION OF MICROBIOLOGICAL CRITERIA – ESSENTIAL CONSIDERATIONS

Selection of a standard, specification, or guideline for a particular application is a complicated process. This fact becomes evident as soon as it is realized that a great diversity exists among food products, their microflora, and operations by which they are processed and prepared. For example, any relationship between the bacterial count of a finished product and the sanitary condition that prevailed during its processing could be rendered meaningless if a severe microbial destructive treatment had been applied near the final stage of processing.

To fulfil the purpose of microbiological criteria the following factors, as stated in the 'General Principles,' should be considered:

1 the evidence of hazard(s) to health;
2 the microbiology of the raw material;
3 the effect of processing on the microbiology of the food;
4 the likelihood and consequence of microbial contamination and/or growth during subsequent handling and storage;
5 the category of consumer at risk;
6 the cost/benefit ratio associated with the application of the criterion.

Ultimately, for microbiological criteria to be meaningful among the criteria for judging food safety and quality the following are required: (a) knowledge of the specific aetiological agents of food-borne disease and their significance, and of other organisms indicative of faulty manufacturing practice; (b) adequate analytical methods and continued work toward development of new and improved methods; and (c) appropriate sampling procedures, and rational bases for sampling and setting the microbiological limits that will be tolerated. The reader will note that the above needs relate directly to the components of a microbiological criterion. The following is a brief summary of current information relative to these needs.

(a) *Aetiological agents or contaminants*

Periodic reports, annual summaries, texts, and reviews dealing with food-borne disease (Anon, 1980, 1981; CDC, 1983; Health and Welfare, Canada, 1984; FAO/WHO, 1974, 1975; Okabe, 1973; Riemann and Bryan, 1979; WHO, 1974, 1976) serve to justify the selection of pathogens associated with food-borne outbreaks that are discussed in Chapter 5. The pathogens are grouped into three categories according to severity of risk. The decision whether or not to include a pathogen in one of these groups depends on its relevance to the microbiological safety of the food.

Salmonella, *Staphylococcus aureus*, *Clostridium perfringens*, and *Vibrio parahaemolyticus* continue to be the most frequent causative agents of food-borne illness. The incidence of outbreaks caused by each varies according to country or geographical location. For example, at least 40–50% of reported outbreaks of food-borne illness occurring in Japan have been due to *V. parahaemolyticus*, almost always from consumption of raw fish. Occasional outbreaks occur in western countries, but the vehicle of transmission is usually processed rather than raw seafoods.

In many western countries the combined annual total number of outbreaks of illness caused by *Salmonella*, *Staph. aureus*, and *C. perfringens* represents 70–80% of the reported outbreaks due to bacteria. Botulism outbreaks, although relatively infrequent, are of serious concern, primarily because of the life-threatening nature of this disease. Commercially processed foods are infrequently involved, home canned or prepared foods being the common vehicle of transmission.

Other pathogens, such as pathogenic *Escherichia coli*, *Shigella*, *Bacillus cereus*, *Yersinia enterocolitica*, *Vibrio cholerae*, *Streptococcus pyogenes*, *Mycobacterium tuberculosis*, *Brucella abortus*, *Corynebacter diphtheriae*, *Campylobacter fetus* subsp. *jejuni*, and viruses, are known causes of food-borne illnesses but either they currently are infrequently involved in outbreaks or the extent and importance of their role has not been adequately assessed.

Often the pathogen responsible for an outbreak of illness is not identified, as in 30–60% of outbreaks reported to the Center for Disease Control (USA) in each of the years 1977–81 (CDC, 1983). This failure of indentification may have been because of late or incomplete laboratory investigation, because the responsible organism was not yet recognized as a cause of food-borne disease, or because it could not be identified by available laboratory techniques.

In most countries the reporting of cases and investigation of outbreaks of food-borne illness is grossly incomplete. In Canada and the USA, for example, the number of cases and outbreaks that actually occur is estimated to be many times greater than the reported figures (Hauschild and Bryan, 1980). Such under-reporting is unfortunate for in addition to providing a continuing assessment of trends in aetiological agents and food vehicles, as indicated above, prompt reporting and investigation of outbreaks provides for (a) identification and removal of contaminated products form the market, (b) correction of faulty food-preparation practices in food-service establishments, processing plants, and homes, (c) identification and appropriate treatment of human carriers of food-borne pathogens, and (d) possible detection of new agents of food-borne disease.

The presence of species and groups of some microorganisms in food may indicate the potential for presence of pathogens or the extent to which Good Manufacturing Practices (GMP) prevailed during manufacture or processing. The relationship between the presence of 'indicator' organisms and either of these is, of course, indirect, and probabilities of positive correlations range from high to zero. An example of the latter occurrence has already been given – that of a bacterial count of an end-product having received a microbiologically destructive treatment late in its processing. High probabilities occur when analytical findings reveal indicator organisms that are known to build up on equipment surfaces at 'critical control points' if faulty cleaning or some other neglectful or faulty practice has occurred. It is essential, therefore, to know which indicator species or group to seek and to include among the components of a microbiological criterion (ICMSF, 1978a, 1978b, 1980). Much of the reported research in food microbiology is also related to this subject.

(b) *Methodology*

Microorganisms in Foods, 1 (ICMSF 1978a) presents methods for the detection and enumeration of the principal food-borne pathogens and of groups or species whose presence in foods (within numerical limits) indicates exposure to conditions that might introduce hazardous organisms and/or allow their proliferation.

For many years the International Dairy Federation has been active in standardizing methods. Several have been published and others are in various stages of development (IDF, 1981–82).

Similarly the International Standards organization (ISO), through several subcommittees of Technical Committee 34 (Agricultural Products), has issued several 'international methods' (ISO, 1983); others are under development. ISO methods are specified where applicable in Codex Codes of Hygienic Practice.

The American Association of Official Analytical Chemists (AOAC) was one of the first to initiate systematic studies to standardize analytical methods for foods. Where applicable, AOAC methods are considered official for national, state, and local government agencies having food-regulatory responsibilities in the United States (AOAC, 1980).

Methods for broader application – that is, not limited to detection and enumeration of pathogens and indicator organisms – are contained in a *Compendium of Methods* published by the American Public Health Association (APHA; Speck, 1984). This publication represents probably the most comprehensive compilation of useful methods in this field. Also, *Standard Methods for the Examination of Dairy Products*, first published in 1910 and now in its 14th edition, is well known and used world-wide (Marth, 1978).

In 1975 joint IDF/ISO/AOAC working groups were established to harmonize methods and standardize work of the three organizations. Through common membership, related work of ICMSF is coordinated with work of these groups, and significant contributions towards minimizing differences in basic provisions in methods of mutual concern have been made. Methods included in the APHA and ICMSF publications referred to above differ from corresponding ISO, IDF, and AOAC methods primarily in format.

Thus, for almost a century research in laboratories throughout the world has steadily progressed in the development of methods for the microbiological examination of foods. As a consequence, some of our present methods are precise and capable of being uniformly applied by well-trained analysts having adequate facilities. However, many methods are complex and time-consuming, which limits their applications. Methods for some purposes are inadequate or produce equivocal results (e.g., for certain of the toxins of enteric pathogens, and for determination of pathogenicity among strains of the same species such as *E. coli*, *Y. enterocolitica*, and *Campylobacter fetus* subsp. *jejuni*). There is need, therefore, for fostering research toward solution of these difficulties and for continued coordination among methods-standardizing organizations. Such efforts will minimize one deterrent to broader application of microbiological criteria: the lack of adequate and standardized methods. No useful purpose would be served by establishment of a criterion without an adequate and practical method either to detect or to enumerate the contaminant of concern.

(c) *Sampling and sampling plans*

Once it is decided to analyse a food for presence of a contaminant(s) one must examine either the whole lot or a sample of it. The former would destroy the lot and thus be impractical. Two questions then arise: what should be the size of the sample? and what conclusions can be drawn from the analytical results? Answering these questions involves statistical concepts of population probability and sampling. These and other related considerations are treated in Part I of this book.

Briefly, a sampling plan is a procedure for withdrawing a sample, carrying out analysis of appropriate sample units, and making appropriate decisions based on an agreed criterion. The criterion may be the absence of a particular contaminant in a proportion or all of the sample units examined (e.g., *Salmonella*) or the number of sample units that must not exceed a specified total bacterial count or count of an indicator species or group.

A sampling plan specified in a microbiological criterion should be consistent with the hazard likely to be associated with the food: the greater the hazard the greater the stringency of the sampling plan. This principle was used by the *Salmonella* Committee (NAS-NRC, 1969) in developing a system for classification of foods into five categories according to the risk of presence of *Salmonella* in food. Classification was based on three product characteristics: (a) the product contains an ingredient identified as a potential factor in salmonellosis, (b) the manufacturing process does not include a process that destroys *Salmonella*, and (c) there is a substantial likelihood of growth of *Salmonella* if the product is mishandled or 'abused' in distribution or consumer usage. Subsequently, ICMSF proposed a similar but somewhat modified system, which provides for 15 hazard categories, called cases (see Chapter 5, Table 6). Sampling plans thought to be appropriate for each hazard category (case) are suggested (see Table 10).

Ultimately, the choice of a sampling plan for a particular food or related product will depend on how well the microbiology of that food is understood. Such understanding on the part of the Commission based upon data from various sources, adherence to 'General Principles,' and, to some extent, expert opinion formed the basis for the applications of microbiological criteria suggested in Part II of this book.

D. INTERNATIONAL DEVELOPMENTS AND APPLICATIONS

Microbiological criteria for foods at the international level find expression within the framework of the joint FAO/WHO food standards program as implemented by the Codex Alimentarius Commission and described in Kermode and McNally (1973) and the *Procedural Manual* of the Codex Alimentarius Commission (Codex, 1975). The program was the direct result of growing conflicts between requirements in national food legislation and the general requirements of the main food markets of the world manifested by serious non-tariff obstacles to trade caused by differing national food legislation. The Commission's objectives are to develop international food standards on a world-wide or sectional basis, as well as other recommended measures, such as codes of practice and guidelines, with the expectation that their general adoption would serve to remove non-tariff barriers to food trade and to avoid them in the future.

Several aspects of food control are covered in Codex standards and codes of practice, including composition, labelling, additives, and hygiene. Subsidiary bodies of the Commission, the Codex committees, develop the standards and codes. While the period from drafting to eventual adoption of a standard or code is lengthy (normally four to five years), the system has worked well as 130 international food standards and codes of practice have been established and others are in various stages of development. The Codex Committee on Food Hygiene has the major responsibility for all provisions on food hygiene in Codex documents, including microbiological criteria. Emphasis has been placed on codes of hygienic practice (effectively good manufacturing practices). The Committee has need for expert advice in dealing with highly specialized microbiological matters and especially in developing microbiological criteria. Assistance has been provided by ICMSF, particularly, through their three publications; further, the ICMSF 2- and 3-class sampling plans have been adapted by the Committee. Also invaluable assistance was provided by a series of Joint FAO/WHO Consultations (FAO/WHO, 1975, 1977, 1979, 1980).

The first Consultation (FAO/WHO, 1975) reviewed previous work on microbiological criteria and recommended that microbiological end-product specifications be included in a 'Code of Hygienic Practice for Egg Products' then being developed by the Committee. This became the first microbiological criterion (specification) in a code of hygienic practice.

A second Consultation (FAO/WHO, 1977) considered microbiological criteria for several other products for which codes of hygienic practice were being developed. Additionally, the Consultation prepared the 'General Principles' document which has been referred to repeatedly above. Other recommendations from the second Consultation included microbiological specifications for precooked shrimps and prawns and for foods for infants and children. These criteria have been incorporated in the respective codes of hygienic practice for these products. Further, the Consultation suggested that criteria be considered for a number of other foods but recommended more immediate attention to dried milk, chilled and frozen raw poultry, chilled and frozen raw meat, cheese, precooked frozen crab meat, desiccated coconut, and precooked frozen lobster.

A third Consultation (FAO/WHO, 1979) met primarily to consider microbiological criteria for chilled and frozen raw meat and poultry. Criteria for these products were deemed inadvisable (see Chapters 11 and 13).

In the most recent consideration of microbiological criteria (FAO/WHO, 1980) the end-product specifications for dried mild products were recommended. These are contained in the current draft of the 'Code of Hygienic Practice for Dried Milk' (Codex, 1981c) and are further discussed in Chapter 15.

In light of the above and the continued interest of ICMSF in microbiological criteria as meaningful among mechanisms for judging food safety and quality, this revision of its second book has been prepared. In setting forth the principles of sampling, and the various sampling plans which are suggested, special effort has been made to be consistent with current knowledge. Accretion of new knowledge about any subject as dynamic as food safety and quality is inevitable. Its careful evaluation and application will determine the extent to which microbiological criteria will continue to serve a useful purpose.

REFERENCES

Anon 1980. Surveillance of food poisoning and salmonella infections in England and Wales, 1970–9. *British Medical Journal*, *281*, 817–818

– 1981. Food poisoning and salmonellosis surveillance in England and Wales, 1980. *British Medical Journal*, *283*, 924–925

AOAC (Association of Official Analytical Chemists). 1980. *Official Methods of Analysis*. AOAC, Washington, DC

CDC (Centers for Disease Control). 1983. *Foodborne Disease Surveillance, Annual Summary, 1981*. CDC, Atlanta, Georgia

Codex (Codex Alimentarius Commission). 1975. *Procedural Manual*, 4th ed. Codex Alimentarius Commission, FAO, Rome

– 1981a. *Codex Alimentarius Commission, Fourteenth Session, 1981: Report of 17th Session of the Codex Committee on Food Hygiene, Washington, D.C. 17–21 November 1980*. ALINORM 81/13. Appendix II, General Principles for the Establishment and Application of Microbiological Criteria for Foods. Codex Alimentarius Commission, FAO, Rome

– 1981b. *Codex Alimentarius Commission, Fourteenth Session, 1981: Report of 17th Session of the Codex Committee on Food Hygiene, Washington, D.C. 17–21 November 1980*. ALINORM 81/13. Appendix V, Extract from the Report of the FAO/WHO Working Group on the Establishment and Application of Microbiological Criteria for Dried Milk Products and Natural Mineral Waters. Codex Alimentarius Commission, FAO, Rome

– 1981c. *Codex Alimentarius Commission, Fourteenth Session, 1981: Report of 17th Session of the Codex Committee on Food Hygiene, Washington, D.C. 17–21 November 1980*. ALINORM 81/13. Appendix IV, Proposed Draft Code of Hygienic Practice for Dried Milk. Codex Alimentarius Commission, FAO, Rome

Corlett, D.A. 1974. Setting microbiological limits in the food industry. *Food Technology*, *28*, 34–40

FAO/WHO (Food and Agricultural Organization / World Health Organization). 1974. *Identification of Contaminants to be Monitored and Recommendations on Sampling Plans and Methodology*. Report of an Expert Consultation on Joint FAO/WHO Food Contamination Monitoring Programme, 7–11 October 1974. WHO, Geneva

– 1975. *Microbiological Specifications for Foods*. Report of a Joint FAO/WHO Expert Consultation, Geneva, 7–11 April 1975 (Document EC/Microbiol/75 Report 1). FAO, Rome

– 1977. *Microbiological Specifications for Foods*. Report of the Second Joint FAO/WHO Expert Consultation, Geneva, 21 February–2 March 1977 (Document EC/Microbiol/77 Report 2). FAO, Rome

– 1979. *Microbiological Criteria for Foods*. Report on a Joint FAO/WHO Working Group on Microbiological Criteria for Foods, Geneva, 20–26 February 1979 (Document WG/Microbiol/79/1). WHO, Geneva

– 1980. *Report of the Fourth Working Group on the Establishment and Application of Microbiological Criteria for Foods – Dried Milk and Natural Mineral Water, Washington, D.C.* (Document VPH/81.32). FAO, Rome

Foster, E.M. 1969. Bacteriological standards for foods – Their significance. *Quarterly Bulletin of the Association of Food and Drug Officials of the United States*, *33*, 119–130

Hauschild, A.H.W., and Bryan, F.L. 1980. Estimate of cases of food- and waterborne illness in Canada and the United States. *Journal of Food Protection*, *43*, 435–440

Health and Welfare Canada. 1984. *Food-Borne and Water-Borne Disease in Canada, Annual Summary 1978*. Issued 1984. Health Protection Branch, Health and Welfare Canada, Ottawa

ICMSF. 1978a. *Microorganisms in Foods. 1. Their Significance and Methods of Enumeration*, 2nd ed. University of Toronto Press, Toronto

– 1978b. *Microorganisms in Foods. 2. Sampling for Microbiological Analysis: Principles and Specific Applications*. University of Toronto Press, Toronto

– 1980. *Microbial Ecology of Foods*. Vol. 2. *Food Commodities*. Academic Press, New York

IDF. 1981–82. *International Dairy Federation Catalogue 1981–82*. International Dairy Federation, Brussels, Belgium

ISO. 1983. *International Standards Organization. Catalogue 1983*. International Organization for Standardization, Geneva, Switzerland

Kermode, G.O., and McNally, H.S. 1973. Joint FAO/WHO Food Standards Programme, Codex Alimentarius Commission. *Nutrition Newsletter* (FAO), 11 (1), 1–17

Kronick, D. 1973. Are microbiological standards for foods needed? *Quarterly Bul-*

letin of the Association of Food and Drug Officials of the United States, *37*, 73–77

Marth, E.H. (ed). 1978. *Standard Methods for the Examination of Dairy Products*, 14th ed. American Public Health Association, Washington, DC

NAS-NRC (National Academy of Sciences–National Research Council). 1964. *An Evaluation of Public Health Hazards from Microbiological Contamination of Foods*. Publication No. 1195, NAS-NRC, Washington, DC

– 1969. *An Evaluation of the Salmonella Problem*. Publication No. 1683, NAS-NRC, Washington, DC

Okabe, S. 1973. Statistical review of food poisoning in Japan – especially that by *Vibrio parahaemolyticus*. In *Proceedings of an International Symposium on Vibrio parahaemolyticus, September 17–18, 1973*, ed. T. Fujino, G. Sakaguchi, R. Sakazaki, and Y. Takeda. Saikon Publishing Co., Tokyo

Olson, J.C. Jr. 1975. Some scientific and regulatory developments of national and international significance in food microbiology. *Journal of the Food Hygienic Society of Japan*, *16*, 352–367

– 1978. Microbiological specifications for foods: International activities. *Food Technology*, *32*, 55–57, 62

Peterson, A.C. 1970. Development of microbiological standards for ready-to-eat foods. *Quarterly Bulletin of the Association of Food and Drug Officials of the United States*, *34*, 114–125

Rieman, H., and Bryan, F.L. (eds). 1979. *Food-Borne Infections and Intoxications*, 2nd ed. Academic Press, New York

Speck, M.S. (ed.) 1984. *Compendium of Methods for the Microbiological Examination of Food*, 2nd ed. American Public Health Association, Washington, DC

Thatcher, F.S. 1955. Microbiological standards for foods: Their function and limitations. *Journal of Applied Bacteriology*, *18*, 449–461

WHO. 1974. World Health Organization. *Foodborne Disease: Methods of Sampling and Examination in Surveillance Programmes*. Technical Report Series No. 543, WHO, Geneva

– 1976. World Health Organization. *Microbiological Aspects of Food Hygiene*. Technical Report Series No. 598, WHO, Geneva

2

Concepts of probability and sampling

Consider a trial or a test the outcome of which is doubtful, such as a test for the presence of an organism in a food. After the standard procedure has been followed, the test either shows the presence of the organism or it does not; thus we have a positive or a negative result. We have resolved the pre-test uncertainty.

If there were relatively many of the organisms present in the food sampled, we would expect relatively many such tests to yield a positive result. But if there were relatively few organisms present, we would expect that relatively few tests would yield a positive result. In the two cases, the 'probability' of a positive result would be respectively high and low.

A. PROBABILITY

The probability of a positive result is, in fact, the 'long-run' proportion of times a positive result occurs out of all the times we test the food. Thus, if a positive result occurs 112 times in 1000 tests, we estimate the probability to be 112/1000 = 0.112, while if it occurs 914 times in 1000 trials, we estimate the probability to be 914/1000 = 0.914.

The word 'estimate' is used because, if we were to run 1000 trials again on the same material and using the same procedure, we could not be sure of observing 112 and 914 positive results, respectively. But the results should be close to these, because 1000 is a large number of trials. Thus the estimated probability of a positive result, or any other outcome in some other type of test, is the proportion of times the outcome did occur among the trials or tests actually made. A probability can be anywhere from 0 to 1. It will be zero if the organism is absent from the food (assuming a test procedure that will detect any organism present), and 1 if every one of a large number of tests provides a positive result.

What does an observed proportion of trials such as 0.112 estimate? Suppose we divided the entire lot of food into small sample units, say perhaps 10,000,000 one-gram units, and then went through the test procedure on each of these sample units. Suppose 1,051,200 gave positive results. Then the ratio 1,051,200/10,000,000 (the actual proportion of positives) = 0.10512 is the measure of the probability. But what kind of measure? This is no longer an 'estimate' of the probability of a positive result. Instead, it is the 'true probability' or 'population probability.' Of course, this approach is not practical, because of the test time needed, and because there would be no food left to eat! But it is useful to have this concept in mind. The 'population probability' determines the kind of sample probabilities or estimated probabilities we may expect from a given number of sample units examined. If that number is low, the estimated probability is not likely to be precise. In general, we never know the true or population probability, but have only estimates of it. But we do know that the larger the number of units included in the sample of the population, the closer the estimated probability is likely to be to the population probability.

B. POPULATION AND SAMPLE OF THE POPULATION

The preceding section introduced the two basic concepts: (i) population as a whole and (ii) sample of the population. In relation to results of an experimental test, these are, respectively, the whole set of results of the test made on each and every unit in the lot of food and the partial collection of such results, derived only from the group of sample units actually examined. In terms of standard plate counts, for example, these concepts would be represented by (i) all the counts that could be made by examining every unit in the lot and (ii) those counts actually made on the few sample units examined. Statisticians use the word 'sample' for the group of units which is withdrawn to estimate the character of a population, while an analyst or bacteriologist would call any one of these units a 'sample.' To try to minimize the confusion, we distinguish between 'the sample of the population' and the 'sample units' of which it is composed.

C. DISTRIBUTION OF COUNTS

Any series of measurements or counts can be analysed by tallying, that is, by taking each number in turn, and tallying it in whatever numerical interval it falls. For example, we might have classes for plate counts of 8000–9999, 10,000–11,999,12,000–13,999, etc. Then we tally each count in the class to which it belongs. This arrangement of observed data into classes is called a 'frequency distribution.' In order to find the estimated probability of a count between, say, 10,000 and 14,000, we add the frequencies for those classes lying between these limits, namely 10,000–11,999 and 12,000–13,999, and then divide the sum of these two frequencies by the total frequency for all the classes. For example, this might be 15 divided by 300, which would be 0.050. In general, the larger the total number of counts tallied, the more the frequencies of the classes tend to smooth out on plotting and appear like a curve.

D. CHOOSING THE SAMPLE UNITS

Chapter 3 describes how to choose the material to be tested from the total amount in the lot or shipment. The important point is to avoid bias, so that the sample of the population may represent the lot as well as possible. Random choice is one way of achieving this. Thus, if we can think of the lot as made up of a population of 10-gram blocks called 'sample units,' and we decide upon a sample from this population of 10 such units, then we should choose these units in such a way that each sample unit in the lot has the same chance of being included among the sample units chosen. At the very least, we should try to draw test material from all parts of the lot.

E. THE SAMPLING PLAN

The sample units so drawn will, after examination, yield results which will be compared with certain criteria to reach a decision as to whether the entire lot should be accepted or rejected (see Section H below for explanation of rejection). The particular choice of sampling procedure and the decision criteria is called the 'sampling plan.' Proper drawing of samples is essential if the decision-making criteria of the sampling plan are to give unbiased results. By sampling randomly we are able to reduce the risk of making wrong decisions.

A simple example of a sampling plan follows. Take 10 sample units of a food from a lot and subject them to appropriate laboratory procedure to test for the presence or absence of a microorganism. If two or fewer of the 10 sample units show presence of the organism, that is, give a 'positive' result, then the whole lot of food is acceptable (relative to this organism). But if three or more give a positive result, the whole lot is to be rejected. This plan is described by $n = 10$ (number of sample units drawn) and $c = 2$ (maximum allowable number of positive results).

F. THE OPERATING CHARACTERISTIC CURVE

In section E we described what we mean by a sampling plan, and gave an example of a simple sampling plan based upon positive and negative indications of a microorganism. This plan was described by the two numbers $n = 10$, $c = 2$. If we are going to use this plan, we want to know what it will do for us. How discriminating is it? It is possible when using such a plan that it will sometimes, though rarely, accept a relatively poor lot; it is also possible, though rarely, for it to reject a relatively good lot. There is no way to avoid some degree of risk in each acceptance and in each rejection, unless we test the entire lot, in which case no edible food will be left. We can make these risks smaller by testing more sample units; that is, by using a larger number of samples (larger n). In fact, we can reduce the risks to any desired level by making n sufficiently large. But in practice we usually have to seek a compromise betwen (a) large n (many sample units) and small risks, and (b) small n (few sample units) and large risks.

Here we often use an 'operating characteristic' curve (abbreviated as OC curve). Such a curve has two scales. The vertical scale gives the probability of acceptance, P_a; that is, P_a is the expected proportion of times that the results will indicate 'pass the lot' out of the number of times a lot of this given quality is sampled for a decision. The horizontal scale shows a measure of lot quality. The commonest measure of lot quality is the true probability or percentage of times that a test on a sample unit from a lot of this quality would show an unsatisfactory result (e.g., a 'positive,' or a count above some number m). This is called p, and might be anywhere between 0 and 100%. This probability is commonly expressed as percentage of 'defectives.' Let us see how the probability runs $n = 10$, $c = 2$ (the calculations are carried out by using the so-called binomial distribution):

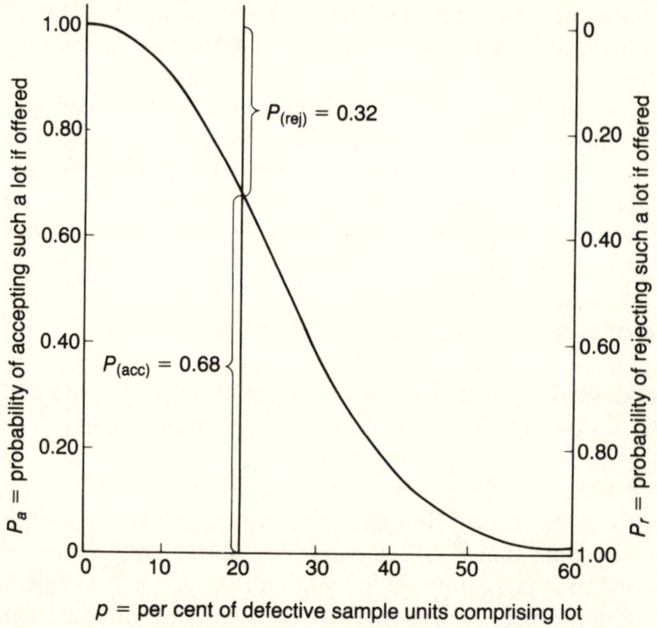

Figure 1. The operating characteristic curve for $n = 10$, $c = 2$, i.e., the probability of accepting lots, in relation to the proportion defective among the sample units comprising the lots

p%	0	10	20	30	40	50	60
P_a	1.00	0.93	0.68	0.38	0.17	0.05	0.01

Figure 1 shows the full curve (cf. Table 3, p. 35). We see that the greater the probability (p) that a test unit from a given lot yields a positive result, the lower is the probability of acceptance (P_a) of that lot. If, for example, we set a limit of 20% defectives (i.e., $p = 20\%$), then $P_a = 0.68$. This means that on 68 of every 100 occasions when we sample a 20% defective lot, we may expect to have 2 or fewer of the 10 tests showing the presence of the organism and thus calling for 'acceptance,' while on 32 of every 100 there will be 3 or more positives, calling for non-acceptance. But if $p = 10\%$ (i.e., 10% positive units in the lot), such lots will be accepted 0.93 of the time; or if $p = 40\%$, such lots will be accepted 0.17 of the time. Thus lots with 10% of the sample units defective will be accepted most of the time, but with 40% defective, rather seldom.

G. CONSUMER RISK AND PRODUCER RISK

Since decisions to accept or reject lots are made on samples drawn from the lots, occasions will arise when the sample results do not reflect the true condition of the lot. The 'producer's risk' describes the probability that an 'acceptable' lot if offered will be falsely rejected. The 'consumer's risk' describes the probability that a 'bad' lot when offered will be falsely accepted. 'Consumer's risk,' for the purpose of this text, is considered to be the probability of accepting a lot whose actual microbial content is substandard as specified in the plan, even though the determined values indicate acceptable quality. This is expressed by the probability of acceptance (P_a) as given in Tables 2–5 in Chapter 4. The 'producer's risk' (the obverse of 'consumer's risk') is expressed by $1 - P_a$.

H. ACCEPTANCE AND REJECTION

We have been speaking of acceptance or rejection of a lot, on the basis of a sampling plan associated with some particular microbiological test. This judgment, of course, applies only to the purpose for which that test (or several such tests) was performed. A food unsuitable for one purpose may still be suitable for another; for example, if 'rejected' for humans it might still be suitable for animals. Or a rejected food might even, if sorted to remove objectionable material or if reprocessed, be so improved as to pass the test and become acceptable for the original purpose. Normally, therefore, a rejected lot will simply be withheld while the responsible authority decides what to do with it: to return it to the producer, order reprocessing, forbid its use for human consumption, or order its destruction, according to circumstances. Throughout this text, the terms 'accept' and 'reject' are used in this limited sense.

3

Principles of drawing samples

Sampling is necessary because it is not feasible to examine microbiologically an entire lot of food. Results from the portion of the lot represented by the sample are used to draw conclusions about the whole (the lot). The process involves the concepts of 'population,' 'sample' (made up of 'sample units'), and the associated 'probability,' discussed in Chapter 2.

A. WHAT IS A 'LOT'?

Ideally a lot is a quantity of food or food units produced and handled under uniform conditions. The implication is that there is a homogeneity within a lot, which, when considering microbial levels and distributions, rarely occurs in practice. In most instances the distribution of microorganisms within 'lots' of food is heterogeneous. Even in food factories this heterogeneity makes the interpretation of sample results difficult, and it becomes even more so in circumstances where a lot is less well defined, such as in a consignment of a very large quantity of food.

It is therefore helpful if suppliers can be persuaded to give identifiable code numbers to batches (lots) of food produced over short time periods (e.g., a day or part of a day), for particular processes. The choice of coding system will vary from process to process depending on the type of process and degree of homogeneity within the batch. For example, a continuous process may produce a relatively homogeneous product, in which case a lot could include units produced continuously over a relatively long time, whereas a batch process, such as batch retorting, may require the coding of relatively few product units as one lot (e.g., one retort load).

If a consignment consisting of a mixture of production batches is treated as one lot, the producer's risk (i.e., the risk that an acceptable lot will be rejected) can be high, even though relatively few of the production batches within the lot are of poor quality. Treating the individual production batches as lots, and coding appropriately, permits more precise identification of poor-quality food and, at the expense of more analyses, can result in the rejection of less units from the whole consignment.

A lot should be composed of food produced with as little variation as possible for a given process or commodity. Because of the uncertainties in identifying a lot commercially, however, the use of the word 'lot' in this book is usually in its 'statistical' sense, as a collection of units of a product, the acceptability of which is determined by examining a sample drawn from it.

B. WHAT IS A 'REPRESENTATIVE' SAMPLE?

A 'representative' sample reflects, as far as is possible, the composition of the lot from which it is drawn.

How then should a representative sample be drawn? It is important to avoid bias and to draw a sufficient number of sample units to confidently make a judgment about a lot. Sampling at random is the universally recognized way of avoiding bias and is more reliable than consciously trying to draw the sample units from the various parts of a lot. The units (cartons or containers, particular weights of solid, or volumes of liquid) are drawn by using random numbers. There is, of course, no guarantee that a random sample (see Section C below) has characteristics identical with those of the lot, but it is more likely to have than samples chosen non-randomly.

It is also possible, and may be desirable, to use a 'stratified' random sampling approach, drawing a random sample of a given number of sample units from each 'stratum' (e.g., each sublot or carton). The objective is to give each small bit of material within each stratum an equal chance to be in the total sample. Such stratification is a device for handling known sources of variation, and may be used where one has prior knowledge that the consignment is potentially not of uniform quality. A consignment might not be of uniform quality if portions of it are shipped in different vehicles or in different holds of a vessel, or if it is known that it is really composed of several lots, representing, for example, different days of production from the same plant or different plants of the same company. The results for different strata should be assessed separately and then pooled if they appear to be homogeneous.

C. USE OF TABLE OF RANDOM NUMBERS

The generally accepted method of random sampling is to use a table of random numbers, comprising columns of single digits 0, 1, 2, ..., 9 in random order (Table 1). These have been generated randomly so that, in any column, any digit is completely independent of the preceding digits in this or other columns. (At each place the probability for each one of the ten digits to appear next is 0.1.)

To use a table of random numbers or digits, first give consecutive numbers to the packages, containers, or other units in the lot that is to be sampled, in any convenient order. Then pick a page of the table of random numbers (Table 1) – for example, by using numbered cards in a box. Next, without looking, bring a pencil point down on some place on the page. The digit nearest the point is the first digit of the number of the first sample package. Take as many columns to the right of that in which this digit appears as are needed to number all the packages in the lot. The digits in these columns in the same row as the first digit chosen complete the number of the first sample. The numbers of additional sample packages are the following numbers in the designated columns. Note that the column numbers are indicated at the top of the table and that each column consists of a vertical row of one-digit numbers. The columns are arranged in groups of five to facilitate identification.

To clarify this procedure, take the following example. Suppose a lot of 600 packages has been numbered 1, 2, ..., 600. Then one needs a random number consisting of three random digits in the chosen three consecutive columns, for each package to be sampled. Suppose the pencil hits row 48 and column 10, on the first page of Table 1. Then in row 48 take the three digits in columns 10, 11, and 12; these are 245. Then take package 245 for the sample. In row 49 in the same three columns one next finds 068. So, take package 68 for the sample. Then go to row 50 on the next double page and find 931 in columns 10, 11, and 12. This is outside the range of 1 to 600, so discard it and also the next number 842. Since the next one is 142, take package 142 for the sample. If selection at random yields a number corresponding to a package that has already been chosen, simply go to the next number, and so on, until the required number of sample packages has been assembled.

D. FUNDAMENTAL PRINCIPLES

The number of sample units n refers to the number of units that are chosen separately and independently. For food in small packages, the sample unit is often the individual package. For bulk materials, it is necessary to imagine a number of sample units within each large container, of a size equal to that subsequently subjected to individual analysis. The concepts of random sampling, stratification, and heterogeneity are relevant (see Section B above). A sample of 1000 g consisting of ten units of 100 g individually collected at random would consist of 10 sample units ($n = 10$), whereas, if all 1000 g are taken from one place, n would equal 1; or, drawing two independent sample units of 500 g, $n = 2$.

The effective sample of the population consists only of those sample units that are actually examined. If ten sample units are drawn, but only three are examined, the sampling plan is one in which $n = 3$, not $n = 10$, and the probability of accepting defective lots is correspondingly increased.

The governing factor is not the proportion of the lot sampled, but rather the total gross size of the random sample of the population (number and size of sample units) and the related acceptance/rejection criteria. A random sample of 100×5 g sample units does very nearly as well for a 500,000 g lot as it does for a lot of 10,000 g. The main reasons for drawing a somewhat smaller sample of the population (i.e., fewer sample units) from a small than from a large lot are economics and the greater importance of the decision on the large lot. Trying to sample a fixed proportion of the lot is not a sound concept.

The concept of frame vs consignment is applied whenever it is possible to take a random sample not from an entire consignment but only from some accessible portion of the consignment, this portion of the consignment being called the 'frame.' Sample units are then chosen randomly from the frame. After analysis the sampling results and conclusions technically apply only to the frame, not the consignment.

E. PRACTICAL CONSIDERATIONS

To use available resources to the best effect, all food cannot receive the same attention. The important foods, and lots, should receive the most time and effort, which means that they should be subjected to the most intensive sampling. What factors govern this decision?

(a) *Hazard*. The most important factor is the hazard involved. How

TABLE 1*

Ten thousand random digits

	00–04	05–09	10–14	15–19	20–24	25–29	30–34	35–39	40–44	45–49
00	88758	66605	33843	43623	62774	25517	09560	41880	85126	60755
01	35661	42832	16240	77410	20686	26656	59698	86241	13152	49187
02	26335	03771	46115	88133	40721	06787	95962	60841	91788	86386
03	60826	74718	56527	29508	91975	13695	25215	72237	06337	73439
04	95044	99896	13763	31764	93970	60987	14692	71039	34165	21297
05	83746	47694	06143	42741	38338	97694	69300	99864	19641	15083
06	27998	42562	63402	10056	81668	48744	08400	83124	19896	18805
07	82685	32323	74625	14510	85927	28017	80588	14756	54937	76379
08	18386	13862	10988	04197	18770	72757	71418	81133	69503	44037
09	21717	13141	22707	68165	58440	19187	08421	23872	03036	34208
10	18446	83052	31842	08634	11887	86070	08464	20565	74390	36541
11	66027	75177	47398	66423	70160	16232	67343	36205	50036	59411
12	51420	96779	54309	87456	78967	79638	68869	49062	02196	55109
13	27045	62626	73159	91149	96509	44204	92237	29969	49315	11804
14	13094	17725	14103	00067	68843	63565	93578	24756	10814	15185
15	92382	62518	17752	53163	63852	44840	02592	88572	03107	90169
16	16215	50809	49326	77232	90155	69955	93892	70445	00906	57002
17	09342	14528	64727	71403	84156	34083	35613	35670	10549	07468
18	38148	79001	03509	79424	39625	73315	18811	86230	99682	82896
19	23689	19997	72382	15247	80205	58090	43804	94548	82693	22799
20	25407	37726	73099	51057	68733	75768	77991	72641	95386	70138
21	25349	69456	19693	85568	93876	18661	69018	10332	83137	88257
22	02322	77491	56095	03055	37738	18216	81781	32245	84081	18436
23	15072	33261	99219	43307	39239	79712	94753	41450	30944	53912
24	27002	31036	85278	74547	84809	36252	09373	69471	15606	77209
25	66181	83316	40386	54316	29505	86032	34563	93204	72973	90760
26	09779	01822	45537	13128	51128	82703	75350	25179	86104	40638
27	10791	07706	87481	26107	24857	27805	42710	63471	08804	23455
28	74833	55767	31312	76611	67389	04691	39687	13596	88730	86850
29	17583	24038	83701	28570	63561	00098	60784	76098	84217	34997
30	45601	46977	39325	09286	41133	34031	94867	11849	75171	57682
31	60683	33112	65995	64203	18070	65437	13624	90896	80945	71987
32	29956	81169	18877	15296	94368	16317	34239	03643	66081	12242
33	91713	84235	75296	69875	62414	05197	66596	13083	46278	73498
34	85704	86588	82837	67822	95963	83021	90732	32661	64751	83903
35	17921	26111	35373	86494	48266	01888	65735	05315	79328	13367
36	13929	71341	80488	89827	48277	07229	71953	16128	65074	28782
37	03248	18880	21667	01311	61806	80201	47889	83052	31029	06023
38	50583	17972	12690	00452	93766	16414	01212	27964	02766	28786
39	10636	46975	09449	45986	34672	46916	63881	83117	53947	95218
40	43896	41278	42205	10425	66560	59967	90139	73563	29875	79033
41	76714	80963	74907	16890	15492	27489	06067	22287	19760	13056
42	22393	46719	02083	62428	45177	57562	49243	31748	64278	05731
43	70942	92042	22776	47761	13503	16037	30875	80754	47491	96012
44	92011	60326	86346	26738	01983	04186	41388	03848	78354	14964
45	66456	00126	45685	67607	70796	04889	98128	13599	93710	23974
46	96292	44348	20898	02227	76512	53185	03057	61375	10760	26889
47	19680	07146	53951	10935	23333	76233	13706	20502	60405	09745
48	67347	51442	24536	60151	05498	64678	87569	65066	17790	55413
49	95888	59255	06898	99137	50871	81625	42223	83303	48694	81953

*From *Principles and Procedures of Statistics* by R.G.D. Steel and J.H. Torrie (McGraw-Hill, 1960). Used with permission of McGraw-Hill Book Company, Inc.

	50–54	55–59	60–64	65–69	70–74	75–79	80–84	85–89	90–94	95–99
00	70896	44520	64720	49898	78088	76740	47460	83150	78905	59870
01	56809	42909	25853	47624	29486	14196	75841	00393	42390	24847
02	66109	84775	07515	49949	61482	91836	48126	80778	21302	24975
03	18071	36263	14053	52526	44347	04923	68100	57805	19521	15345
04	98732	15120	91754	12657	74675	78500	01247	49719	47635	55514
05	36075	83967	22268	77971	31169	68584	21336	72541	66959	39708
06	04110	45061	78062	18911	27855	09419	56459	00695	70323	04538
07	75658	58509	24479	10202	13150	95946	55087	38398	18718	95561
08	87403	19142	27208	35149	34889	27003	14181	44813	17784	41036
09	00005	52142	65021	64438	69610	12154	98422	65320	79996	01935
10	43674	47103	48614	70823	78252	82403	93424	05236	54588	27757
11	68597	68874	35567	98463	99671	05634	81533	47406	17228	44455
12	91874	70208	06308	40719	02772	69589	79936	07514	44950	35190
13	73854	19470	53014	29375	62256	77488	74388	53949	49607	19816
14	65926	34117	55344	68155	38099	56009	03513	05926	35584	42328
15	40005	35246	49440	40295	44390	83043	26090	80201	02934	49260
16	46686	29890	14821	69783	34733	11803	64845	32065	14527	38702
17	02717	61518	39585	72863	50707	96115	07416	05041	36756	61065
18	17048	22281	35573	28944	96889	51823	57268	03866	27658	91950
19	75304	53248	42151	93928	17343	88322	28683	11252	10355	65175
20	97844	62947	62230	30500	92816	85232	27222	91701	11057	83257
21	07611	71163	82212	20653	21499	51496	40715	78952	33029	64207
22	47744	04603	44522	62783	39347	72310	41460	31052	40814	94297
23	54293	43576	88116	67416	34908	15238	40561	73940	56850	31078
24	67556	93979	73363	00300	11217	74405	18937	79000	68834	48307
25	86581	73041	95809	73986	49408	53316	90841	73808	53421	82315
26	28020	86282	83365	76600	11261	74354	20968	60770	12141	09539
27	42578	32471	37840	30872	75074	79027	57813	62831	54715	26693
28	47290	15997	86163	10571	81911	92124	92971	80860	41012	58666
29	24856	63911	13221	77028	06573	33667	30732	47280	12926	67276
30	16352	24836	60799	76281	83402	44709	78930	82969	84468	36910
31	89060	79852	97854	28324	39638	86936	06702	74304	39873	19496
32	07637	30412	04921	26471	09605	07355	20466	49793	40539	21077
33	37711	47786	37468	31963	16908	50283	80884	08252	72655	58926
34	82994	53232	58202	73318	62471	49650	15888	73370	98748	69181
35	31722	67288	12110	04776	15168	68862	92347	90789	66961	04162
36	93819	78050	19364	38037	25706	90879	05215	00260	14426	88207
37	65557	24496	04713	23688	26623	41356	47049	60676	72236	01214
38	88001	91382	05129	36041	10357	55558	89979	58061	28957	10701
39	96648	70303	18191	62404	26558	92804	15415	02865	52449	78509
40	04118	51573	59356	02426	35010	37104	98316	44602	96478	08433
41	19317	27753	39431	26996	04465	69695	61374	06317	42225	62025
42	37182	91221	17307	68507	85725	81898	22588	22241	80337	89033
43	82990	03607	29560	60413	59743	75000	03806	13741	79671	25416
44	97294	21991	11217	98087	79124	52275	31088	32085	23089	21498
45	86771	69504	13345	42544	59616	07867	78717	82840	74669	21515
46	26046	55559	12200	95106	56496	76662	44880	98457	84209	01332
47	39698	05999	92290	79024	70271	93352	90272	94495	26842	54477
48	83265	89573	01437	43786	52986	49041	17952	35035	88985	84671
49	15128	35791	11296	45319	06330	82027	90808	54351	43091	30387

TABLE 1 (continued)

	00–04	05–09	10–14	15–19	20–24	25–29	30–34	35–39	40–44	45–49
50	54441	64681	93190	00993	62130	44484	46293	60717	52039	76319
51	08573	52937	84274	95106	89117	65849	41356	65549	78787	50442
52	81067	68052	14270	19718	88499	63303	13533	91882	51136	60828
53	39737	58891	75278	98046	52284	40164	72442	77824	72900	14886
54	34958	76090	08827	61623	31114	86952	83645	91786	29633	78294
55	61417	72424	92626	71952	69709	81259	58472	43409	84454	88648
56	99187	14149	57474	32264	85424	90378	34682	47606	89295	02420
57	13130	13064	36485	48133	35319	05720	76317	70953	50823	06793
58	65563	11831	82402	46929	91446	72037	17205	89600	59084	55718
59	28737	49502	06060	52100	43704	50839	22538	56768	83467	19313
60	50353	74022	59767	49927	45882	74099	18758	57510	58560	07050
61	65208	96466	29917	22862	69972	35178	32911	08172	06277	62795
62	21323	38148	26696	81741	25131	20087	67452	19670	35898	50636
63	67875	29831	59330	46570	69768	36671	01031	95995	68417	68665
64	82631	26260	86554	31881	70512	37899	38851	40568	54284	24056
65	91989	39633	59039	12526	37730	68848	71399	28513	69018	10289
66	12950	31418	93425	69756	34036	55097	97241	92480	49745	42461
67	00328	27427	95474	97217	05034	26676	49629	13594	50525	13485
68	63986	16698	82804	04524	39919	32381	67488	05223	89537	59490
69	55775	75005	57912	20977	35722	51931	89565	77579	93085	06467
70	24761	56877	56357	78809	40748	69727	56652	12462	40528	75269
71	43820	80926	26795	57553	28319	25376	51795	26123	51102	89853
72	66669	02880	02987	33615	54206	20013	75872	88678	17726	60640
73	49944	66725	19779	50416	42800	71733	82052	28504	15593	51799
74	71003	87598	61296	95019	21568	86134	66096	65403	47166	78638
75	52715	04593	69484	93411	38046	13000	04293	60830	03914	75357
76	21998	31729	89963	11573	49442	69467	40265	56066	36024	25705
77	58970	96827	18377	31564	23555	86338	79250	43168	96929	97732
78	67592	59149	42554	42719	13553	48560	81167	10747	92552	19867
79	18298	18429	09357	96436	11237	88039	81020	00428	75731	37779
80	88420	28841	42628	84647	59024	52032	31251	72017	43875	48320
81	07627	88424	23381	29680	14027	75905	27037	22113	77873	78711
82	37917	93581	04979	21041	95252	62450	05937	81670	44894	47262
83	14783	95119	68464	08726	74818	91700	05961	23554	74649	50540
84	05378	32640	64562	15303	13168	23189	88198	63617	58566	56047
85	19640	96709	22047	07825	40583	99500	39989	96593	32254	37158
86	20514	11081	51131	56469	33947	77703	35679	45774	06776	67062
87	96763	56249	81243	62416	84451	14696	38195	70435	45948	67690
88	49439	61075	31558	59740	52759	55323	95226	01385	20158	54054
89	16294	50548	71317	32168	86071	47314	65393	56367	46910	51269
90	31381	94301	79273	32843	05862	36211	93960	00671	67631	23952
91	98032	87203	03227	66021	99666	98368	39222	36056	81992	20121
92	40700	31826	94774	11366	81391	33602	69608	84119	93204	26825
93	68692	66849	29366	77540	14978	06508	10824	65416	23629	63029
94	19047	10784	19607	20296	31804	72984	60060	50353	23260	58909
95	82867	69266	50733	62630	00956	61500	89913	30049	82321	62367
96	26528	28928	52600	72997	80943	04084	86662	90025	14360	64867
97	51166	00607	49962	30724	81707	14548	25844	47336	57492	02207
98	97245	15440	55182	15368	85136	98869	33712	95152	50973	98658
99	54998	88830	95639	45104	72676	28220	82576	57381	34438	24565

SOURCE: Prepared by Fred Gruenberger, Numerical Analysis Laboratory, University of Wisconsin, Madison, Wis., 1952

	50–54	55–59	60–64	65–69	70–74	75–79	80–84	85–89	90–94	95–99
50	58649	85086	16502	97541	76611	94229	34987	86718	87208	05426
51	97306	52449	55596	66739	36525	97563	29469	31235	79276	10831
52	09942	79344	78160	11015	55777	22047	57615	15717	82639	36578
53	83842	28631	74893	47911	92170	38181	30416	54860	44120	73031
54	73778	30395	20163	76111	13712	33449	99224	18206	51418	70006
55	88381	56550	47467	59663	61117	39716	32927	06168	06217	45477
56	31044	21404	15968	21357	30772	81482	38807	67231	84283	63552
57	00909	63837	91328	81106	11740	50193	86806	21931	18054	49601
58	69882	37028	41732	37425	80832	03320	20690	32653	90145	03029
59	26059	78324	22501	73825	16927	31545	15695	74216	98372	28547
60	38573	98078	38982	33078	93524	45606	53463	20391	81637	37269
61	70624	00063	81455	16924	12848	23801	55481	78978	26795	10553
62	49806	23976	05640	29804	38988	25024	76951	02341	63219	75864
63	05461	67523	48316	14613	08541	35231	38312	14969	67279	50502
64	76582	62153	53801	51219	30424	32599	49099	83959	68408	20147
65	16660	80470	75062	75588	24384	27874	20018	11428	32265	07692
66	60166	42424	97470	88451	81270	80070	72959	26220	59939	31127
67	28953	03272	31460	41691	57736	72052	22762	96323	27616	53123
68	47536	86439	95210	96386	38704	15484	07426	70675	06888	81203
69	73457	26657	36983	72410	30244	97711	25652	09373	66218	64077
70	11190	66193	66287	09116	48140	37669	02932	50799	17255	06181
71	57062	78964	44455	14036	36098	40773	11688	33150	07459	36127
72	99624	67254	67302	18991	97687	54099	94884	42283	63258	50651
73	97521	83669	85968	16135	30133	51312	17831	75016	80278	68953
74	40273	04838	13661	64757	17461	78085	60094	27010	80945	66439
75	57260	06176	49963	29760	69546	61336	39429	41985	18572	98128
76	03451	47098	63495	71227	79304	29753	99131	18419	71791	81515
77	62331	20492	15393	84270	24396	32962	21632	92965	38670	44923
78	32290	51079	06512	38806	93327	80086	19088	59887	98416	24918
79	28014	80428	92853	31333	32648	16734	43418	90124	15086	48444
80	18950	16091	29543	65817	07002	73115	94115	20271	50250	25061
81	17403	69503	01866	13049	07263	13039	83844	80143	39048	62654
82	27999	50489	66613	21843	71746	65868	16208	46781	93402	12323
83	87076	53174	12165	84495	47947	60706	64034	31635	65169	93070
84	89044	45974	14524	46906	26052	51851	84197	61694	57429	63395
85	98048	64400	24705	75711	36232	57624	41424	77366	52790	84705
86	09345	12956	49770	80311	32319	48238	16952	92088	51222	82865
87	07086	77628	76195	47584	62411	40397	71857	54823	26536	56792
88	93128	25657	46872	11206	06831	87944	97914	64670	45760	34353
89	85137	70964	29947	27795	25547	37682	96105	26848	09389	64326
90	32798	39024	13814	98546	46585	84108	74603	94812	73968	68766
91	62496	26371	89880	52078	47781	95260	83464	65942	91761	53727
92	62707	81825	40987	97656	89714	52177	23778	07482	91678	40128
93	05500	28982	86124	19554	80818	94935	61924	31828	79369	23507
94	79476	31445	59498	85132	24582	26024	24002	63718	79164	43556
95	10653	29954	97568	91541	33139	84525	72271	02546	64818	14381
96	30524	06495	00886	40666	68574	49574	19705	16429	90981	08103
97	69050	22019	74066	14500	14506	06423	38332	34191	82663	85323
98	27908	78802	63446	07674	98871	63831	72449	42705	26513	19883
99	64520	16618	47409	19574	78136	46047	01277	79146	95759	36781

hazardous is the type (or types) or microorganisms present? and how hazardous are the numbers present? These questions are discussed more fully in Chapter 6. The probability of accepting a lot which should be rejected should be reduced with increasing hazard.

(b) *Uniformity*. If the food is well mixed as a result of the manufacturing process (e.g., a mobile liquid), relatively few sample units can be used. If, however, there is likely to be relatively heavy contamination in a few small places in the lot of food, relatively large numbers of sample units may be needed to identify it.

(c) *Stratification*. If it is known that there is stratification within lots of the food (see Section B above), a corresponding stratification can be used in selecting sampling units.

(d) *Record of consistency*. A consistently good record for a food from a specific source indicates that its control is good, and reliable, and hence may justify reduced sampling or even omission of sampling on occasional lots. Discretion must be used in making this decision, with increasing confidence as the record accumulates. Confidence will be maintained if the producer provides detailed records of control procedures.

(e) *Practical limitations*. Since regulatory agencies never have the resources to test all imported lots, they must reduce the sampling plans to a feasible level. Most microbiological tests are laborious and slow; yet regulatory agencies cannot hold highly perishable foods pending results of analysis. Political or administrative pressures to reduce sampling will increase the probability of error.

In considering the above factors, it must be remembered that the ability to distinguish between good and bad lots often improves relatively little with increase in the number of sample units constituting the population sample. Indeed, in many instances reliability only increases roughly as the square root of the number of sample units, so that multiplying the sample number by four will only halve the likelihood of making wrong decisions.

4

Appropriate sampling plans

This text is concerned primarily with plans that may be applied to lots of food presented for acceptance at ports or other points of entry. Often little or no information is available to the receiving agency about the method by which the food was processed or the record of previous performance by the same processor. Under these circumstances attributes plans are appropriate. Variables sampling plans, which depend upon the nature of the frequency distribution of organisms within lots of foods, are suitable only if this distribution is known. This limits severely their usefulness in port-of-entry sampling, but they may be particularly helpful to food producers monitoring their own production.

A. TWO-CLASS ATTRIBUTES PLANS

A simple way to decide whether to accept or reject a food lot may be based on some microbiological test performed on several (n) sample units. This will usually be a test for the presence (positive result) or absence (negative result) of an organism. Concentrations of micro-organisms can be assigned to a particular attribute class by determining whether they are above (positive) or below (negative) some preset concentration.

As explained in Chapter 2, the decision-making process is defined by two numbers. The first is represented by the letter n, and defines the number of sample units required for testing. The second number, denoted c, is the maximum allowable number of sample units yielding unsatisfactory test results: for examples, the presence of the organism, or a count above the defined concentration denoted by m, which in a 2-class plan separates good quality from defective quality. Thus, in a presence/absence decision on a lot, the sampling plan $n = 10$, $c = 2$

means that 10 sample units are taken and a test made on each; if 2 or fewer show the presence of the organisms, the lot is accepted (with respect to this characteristic); but if 3 or more of the 10 show the presence of the organism, the lot is rejected, although not necessarily destroyed (see Chapter 2, Section H, for an explanation of rejection).

The stringency of the sampling plan depends upon *n* and *c* (see Chapter 2, Section E). The larger the value of *n* at a given value of *c*, the better the food quality must be to have the same chance of being passed. Conversely, for a given sample size *n*, if *c* is increased the plan becomes more lenient and will more often pass food lots of a given quality whenever offered for test (i.e., there is a higher probability of acceptance, P_a). Thus, compared with $n = 10$, $c = 2$, the plans $n = 15$, $c = 2$ and $n = 10$, $c = 1$ are more stringent, while the plans $n = 5$, $c = 2$ and $n = 10$, $c = 3$ are more lenient. Probabilities of acceptance for a set of plans are given in Tables 2 and 3. Figure 2 displays the operating characteristic (OC) curves for a few of these plans to illustrate the characteristics of various 2-class attributes plans.

In foods that are intimately mixed, it is possible that the distribution of microorganisms within them may approach uniformity. Under those circumstances, data obtained from the application of a 2-class plan based on presence/absence of an organism may give an approximate guide to the likely concentration of that microbe within the lot. Such techniques should only be applied when the analyst is confident that there is a uniform distribution of the microorganisms of concern.

B. THREE-CLASS ATTRIBUTES PLANS

Three-class attributes plans (see Bray, Lyon, and Burr, 1973) were devised for the situation where the quality of the product can be divided into three attribute classes depending upon the concentration of microorganisms within the sample units. A count above a concentration *M* for any sample unit is unacceptable, and if any count for the *n* sample units from a lot exceeds *M*, that lot is rejected. Counts above *m*, which in a 3-class plan separates good quality from marginally acceptable quality, and up to and including *M* are undesirable but some can be accepted, the number acceptable being denoted by *c*.

Accordingly, in the 3-class plans there are again only two numbers, *n* and *c*, from which it is possible to find the probability of acceptance,

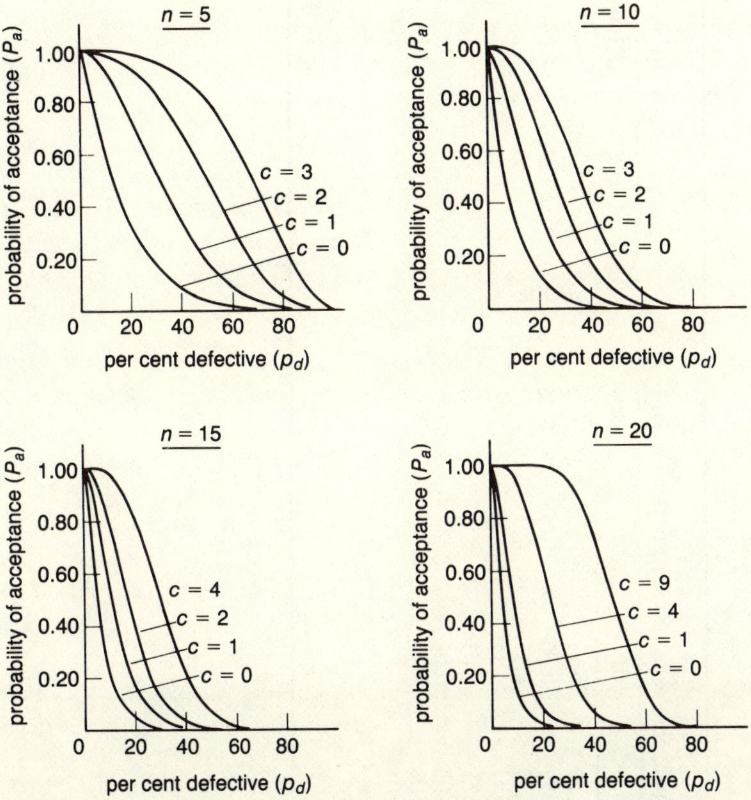

Figure 2 Operating characteristic curves for different sample sizes (*n*) and different criteria of acceptance (*c*) for 2-class attributes plan

P_a, for a food lot of given microbiological quality. To describe the lot quality we consider all sample units which could be drawn from the lot, which must yield counts in three classes; 0 to *m*, *m* to *M*, and above *M*. Since the proportions in the lot for the three classes must total 1, one need only specify *two* of them in describing lot quality. We might call these proportions the per cent 'defective,' i.e. above *M* (P_d), and the per cent 'marginally acceptable,' i.e. *m* to *M* (P_m). The per cent 'acceptable,' equal to or less than *m*, must be 100% minus the sum of P_d and P_m. By appropriate calculations, we can find the probability of acceptance, P_a, for a given lot quality for any specified sampling plan. For example, for the plan $n = 10$, $c = 2$, P_a will be

TABLE 2

Two-class plans ($c = 0$): probabilities of acceptance (P_a) of lots containing indicated proportions of acceptable and defective sample units*

Composition of lot		Number of sample units tested from the population (n)							
% acceptable ($100 - p$)	% defective (p)	3	5	10	15	20	30	60	100
98	2	0.94	0.90	0.82	0.74	0.67	0.55	0.30	0.13
95	5	0.86	0.77	0.60	0.46	0.36	0.21	0.05	0.01
90	10	0.73	0.59	0.35	0.21	0.12	0.04	<	<
80	20	0.51	0.33	0.11	0.04	0.01	<		
70	30	0.34	0.17	0.03	<	<			
60	40	0.22	0.08	0.01					
50	50	0.13	0.03	<					
40	60	0.06	0.01						
30	70	0.03	<						
20	80	0.01							
10	90	<							

* '<' means $P_a < 0.005$.

TABLE 3

Two-class plans (selected c values): probabilities of acceptance (P_a) of lots containing indicated proportions of acceptable and defective sample units*

Composition of lot		$n=5$			$n=10$			$n=15$			$n=20$		
% acceptable $(100-p)$	% defective (p)	$c=3$	$c=2$	$c=1$	$c=3$	$c=2$	$c=1$	$c=4$	$c=2$	$c=1$	$c=9$	$c=4$	$c=1$
98	2	1.00	1.00	1.00	1.00	1.00	0.98	1.00	1.00	0.96	1.00	1.00	0.94
95	5	1.00	1.00	0.98	1.00	0.99	0.91	1.00	0.96	0.83	1.00	1.00	0.74
90	10	1.00	0.99	0.92	0.99	0.93	0.74	0.99	0.82	0.55	1.00	0.96	0.39
80	20	0.99	0.94	0.74	0.88	0.68	0.38	0.84	0.40	0.17	1.00	0.63	0.07
70	30	0.97	0.84	0.53	0.65	0.38	0.15	0.52	0.13	0.04	0.95	0.24	0.01
60	40	0.91	0.68	0.34	0.38	0.17	0.05	0.22	0.03	0.01	0.76	0.05	<
50	50	0.81	0.50	0.19	0.17	0.05	0.01	0.06	<	<	0.41	0.01	<
40	60	0.66	0.32	0.09	0.05	0.01	<	0.01			0.13	<	
30	70	0.47	0.16	0.03	0.01	<		<			0.02		
20	80	0.26	0.06	0.01	<						<		
10	90	0.08	0.01	<									
5	95	0.02	<										

* n = number of sample units tested from the population; '<' means $P_a < 0.005$.

0.21 for a lot distribution for which 20% of the sample counts are marginally acceptable (P_m = 20%) and 10% 'defective' (P_d = 10%). That is, on the basis of the particular values decided upon for *m* and *M*, only about 21 lots out of 100 of that quality will be accepted, because they have no defective counts and two or fewer 'marginally acceptable' counts out of the 10 samples chosen from the lot. The other lots will all be rejected.

Probabilities associated with a collection of 3-class plans are shown in Table 4 for various lot qualities. The scheme depends for acceptance or rejection not only on the proportion of 'defective' material (P_d) but also on the proportion of 'marginally acceptable' product (P_m); that is, on the frequency distribution of microorganisms within the lot. For example, a lot with 20% of 'marginal quality' and 10% 'defective' quality with a sampling plan based on n = 10, c = 2 will, on average, be accepted 21% of the times it is put forward for testing. Using the identical plan (n = 10, c = 2) a lot containing 0% 'defective' material but 40% of 'marginal quality' has a lower chance of acceptance (17%). Thus the 3-class attributes scheme is affected to some extent by the frequency distribution of microorganisms within the batch, but the advantages of the scheme are its simplicity and general applicability, which make it particularly relevant to port-of-entry sampling.

C. COMPARISON OF TWO- AND THREE-CLASS PLANS

Table 5 compares the operating characteristics of the two types of plans recommended in this text, on the basis of equal sample sizes *n*, acceptance numbers *c*, and lot qualities. To facilitate comparison, lot quality is measured as the proportion of the lot worse than level *m* and, correspondingly, the same value of *c* is taken for the 2-class plan as the *c* of marginally acceptable quality for the 3-class plan.

The 2-class plans do not distinguish values between *m* and *M* from those above *M*. If not more than *c* sample units give results above *m*, the lot is acceptable, regardless of how far individual results exceed *m*. But the corresponding 3-class plan does make a distinction, by including an extra subdivision of lot quality since the limit *M* separates marginally acceptable from defective quality.

D. THE INFLUENCE OF SIZE OF LOT

A sample size of, say, n = 30 sample units might be taken at random from a lot of any size (or consignment if appropriate – see Chapter 3,

Section A) for use in a 2-class plan with $c = 0$. If the lot contains a very large number of sample units, one obtains an OC curve such as that shown in Figure 5 (p. 82). There it is apparent that a lot with one defective unit out of 40 (proportion defective, $p = 0.025$) will be accepted about half the time ($P_a = 0.50$) when using the plan $n = 30$, $c = 0$. If, however, the lot contains a smaller number of sample units, the probabilities of acceptance P_a for various values of p would be only slightly smaller than those shown in Figure 5, despite the fact that the 30 sample units might now constitute an appreciable fraction of the whole lot. This effect is only really important when one begins to take a quarter to a half of the lot as a sample, a condition rarely occurring in bacteriological analysis of food lots.

Suppose that instead of $n = 30$, $c = 0$, we use $n = 5$, $c = 0$. The latter plan is quite undiscriminating. But the associated probabilities of acceptance (see Figure 2, p. 33) are again nearly independent of the consignment size.

In summary, a sampling plan of the form described gives nearly the same degree of protection against acceptance of 'bad' lots as against rejection of 'good' ones, regardless of lot size.

The attributes schemes given in this book do not increase stringency as lot size increases. The sampling plan is therefore independent of lot size if the lot is large in comparison to sample size. If the important criterion is the actual number of defective units, rather than the proportion of the batch that is defective, then the sampling plans given are not appropriate.

E. VARIABLES SAMPLING

Under certain circumstances, it has been shown that the logarithms of the counts from a batch of food are likely to be normally distributed (Baird-Parker, 1980; Kilsby and Baird-Parker, 1983 [their Table 2]; Ingram and Roberts, 1976). Knowledge of this 'log-normal' distribution would rarely exist in port-of-entry sampling. For this reason it is not dealt with to any great degree in this book. However, such knowledge may be available to a food producer, particularly when the production process is stable and the microbiological distribution has been carefully studied. Under these circumstances, variables plans are an alternative to attributes plans. Since variables plans may be of use to the food producer in his internal quality control program, they are discussed briefly in Chapter 8.

TABLE 4

Three-class plans: probabilities of acceptance (P_a) of lots containing indicated proportions for selected numbers of sample units and c values (P_d = per cent defective, P_m = per cent marginal)*

	P_m								
P_d	10	20	30	40	50	60	70	80	90
$n=5$, $c=3$									
50	0.03	0.03	0.02	0.01	<				
40	0.08	0.07	0.06	0.04	0.02	<			
30	0.17	0.16	0.15	0.12	0.07	0.03	<		
20	0.33	0.32	0.31	0.27	0.20	0.12	0.04	<	
10	0.59	0.58	0.56	0.52	0.43	0.32	0.18	0.06	<
5	0.77	0.77	0.75	0.69	0.60	0.47	0.31	0.14	0.02
0	1.00	0.99	0.97	0.91	0.81	0.66	0.47	0.26	0.08
$n=5$, $c=2$									
50	0.03	0.02	0.01	<					
40	0.08	0.06	0.04	0.02	<				
30	0.16	0.14	0.11	0.06	0.02	<			
20	0.32	0.29	0.24	0.16	0.09	0.03	0.01	<	
10	0.58	0.55	0.47	0.36	0.23	0.12	0.05	0.01	<
5	0.77	0.72	0.63	0.50	0.35	0.20	0.09	0.02	<
0	0.99	0.94	0.84	0.68	0.50	0.32	0.16	0.06	0.01
$n=5$, $c=1$									
50	0.02	0.01	<						
40	0.06	0.04	0.01	<					
30	0.14	0.09	0.05	0.02	<				
20	0.29	0.21	0.13	0.06	0.02	0.01	<		
10	0.53	0.41	0.27	0.16	0.07	0.03	0.01	<	
5	0.70	0.55	0.38	0.23	0.12	0.05	0.01	<	
0	0.92	0.74	0.53	0.34	0.19	0.09	0.03	0.01	

TABLE 4 (concluded)

P_d	P_m 10	20	30	40	50	60	70	80	90
$n=10,\ c=3$									
40	0.01	<							
30	0.03	0.02	0.01	<					
20	0.10	0.08	0.05	0.02	<				
10	0.34	0.29	0.20	0.10	0.03	0.01	<		
5	0.59	0.51	0.36	0.20	0.08	0.02	<		
0	0.99	0.88	0.65	0.38	0.17	0.05	0.01	<	
$n=10,\ c=2$									
30	0.02	0.01	<						
20	0.09	0.06	0.02	0.01	<				
10	0.32	0.21	0.10	0.04	0.01	<			
5	0.55	0.39	0.20	0.08	0.02	<			
0	0.93	0.68	0.38	0.17	0.05	0.01			
$n=10,\ c=1$									
30	0.02	<							
20	0.07	0.03	0.01	<					
10	0.24	0.11	0.04	0.01	<				
5	0.43	0.21	0.08	0.02	<				
0	0.74	0.38	0.15	0.05	0.01	<			

* Each of these blocks of numbers, relating P_a to P_m and P_d, represents a three-dimensional relation called an OC surface, corresponding to the two-dimensional OC curve.
'<' means $P_a < 0.005$.

TABLE 5

Probabilities of acceptance (P_a) for 2- and 3-class attributes sampling plans when the lot examined contains 5%, 20%, or 50% of sample units > m*

Sampling plan		5% > m			20% > m				50% > m			
		2-class	3-class plan		2-class	3-class plan			2-class	3-class plan		
n	c	plan	0% > M	5% > M	plan	0% > M	10% > M	20% > M	plan	0% > M	25% > M	50% > M
5	0	0.77	0.77	0.77	0.33	0.33	0.33	0.33	0.03	0.03	0.03	0.03
5	1	0.98	0.98	0.77	0.74	0.74	0.53	0.33	0.19	0.19	0.11	0.03
5	2	1.00	1.00	0.77	0.94	0.94	0.58	0.33	0.50	0.50	0.19	0.03
5	3	1.00	1.00	0.77	0.99	0.99	0.59	0.33	0.81	0.81	0.23	0.03
10	0	0.60	0.60	0.60	0.11	0.11	0.11	0.11	<	<	<	<
10	1	0.91	0.91	0.60	0.38	0.38	0.24	0.11	0.01	0.01	0.01	<
10	2	0.99	0.99	0.60	0.68	0.68	0.32	0.11	0.05	0.05	0.02	<
10	3	1.00	1.00	0.60	0.88	0.88	0.34	0.11	0.17	0.17	0.03	<
15	0	0.46	0.46	0.46	0.04	0.04	0.04	0.04	<	<	<	<
20	0	0.36	0.36	0.36	0.01	0.01	0.01	0.01	<	<	<	<
60	0	0.05	0.05	0.05	<	<	<	<	<	<	<	<

* Note that any % > M is included in the % > m; e.g., for the case where 20% > m and 10% > M, it follows that 10% is between m and M.
'<' means $P_a < 0.005$.

REFERENCES

Baird-Parker, A.C. 1980. Food microbiology in the 1980s. *Food Technology in Australia*, *32*, 70–77

Bray, D.F., Lyon, D.A., and Burr, I.W. 1973. Three-class attributes plans in acceptance sampling. *Technometrics*, *15* (3), 575–585

Ingram, M., and Roberts, T.A. 1976. The microbiology of the red meat carcass and the slaughterhouse. *Royal Society of Health Journal*, *96*, 270–276

Kilsby, D.C., and Baird-Parker, A.C. 1983. Sampling programmes for the microbiological analysis of food. In *Food Microbiology, Advances and Prospects*, ed. T.A. Roberts and F.A. Skinner, 307–315. Society for Applied Bacteriology Symposium Series No. 11. Academic Press, London

5

Choice of sampling plan according to purpose

A. GENERAL PRINCIPLES

The stringency of sampling plans for foods is based either on the hazard to the consumer from pathogenic microorganisms and their toxins or toxic metabolites, or on the potential for quality deterioration to an unacceptable state; and it should take account of the types of microorganisms present and their numbers. Some microorganisms merely spoil a product, others can cause illness, and others are taken to indicate the likelihood of contamination by pathogens. Some pathogens cause mild illnesses which seldom spread; others cause mild illnesses which may spread rapidly; and yet others cause severe illness. The degree of food-borne hazard is often increased if the organism has grown to high numbers in the food and, conversely, is usually reduced if the number is reduced. In some cases the food merely acts as a vehicle for transmission of the infectious microorganism. Treatment in the normal course of distribution, storage, and preparation for consumption may decrease, leave unchanged, or increase those numbers, while labile toxins would decompose and stable toxins remain.

The choice of a plan must therefore consider: (i) the type and seriousness of hazards implied by the microorganisms for which the test is to be made; and (ii) the conditions under which the food is expected to be handled and consumed after sampling. Table 6 classifies 15 different cases of sampling plans on a two-dimensional grid taking into account these factors. In the table the stringency of the sampling plan increases with the type and degree of hazard: from a situation of no health hazard but of utility only, through a low indirect health risk (as implied by the presence of indicator organisms), to direct health risks related to disease of moderate or severe implication. The strin-

TABLE 6

Plan stringency (case) in relation to degree of health hazard and conditions of use

Type of hazard	Conditions in which food is expected to be handled and consumed after sampling, in the usual course of events		
	Reduce degree of hazard	Cause no change in hazard	May increase hazard
No direct health hazard			
Utility (e.g., general contamination, reduced shelf-life, and spoilage)	Case 1	Case 2	Case 3
Health hazard			
Low, indirect (indicator)	Case 4	Case 5	Case 6
Moderate, direct, limited spread*	Case 7	Case 8	Case 9
Moderate, direct, potentially extensive spread*	Case 10	Case 11	Case 12
Severe, direct	Case 13	Case 14	Case 15

* See Section B(e) below for explanation of extensive and limited spread.

gency of the sampling plan also changes according to the conditions under which the food is expected to be handled. Hazards may remain unchanged, be reduced by cooking, or increase because of subsequent growth of microorganisms. The most lenient plan is case 1. Stringency increases from left to right and from top to bottom of the table, so that case 15 is the most stringent.

On the basis of the above classification into cases we can now examine in more detail the considerations involved in assessing the type of hazard associated with particular microbial species, in deciding whether likely conditions of use will affect the degree of hazard, in recognizing the corresponding case, and in choosing an appropriate sampling plan.

B. FACTORS THAT INFLUENCE THE TYPE OF HAZARD

Among the many factors that determine the nature of the hazard from a species causing food-borne disease are epidemiological considerations (e.g., frequency of causing disease), clinical features (e.g., severity and duration of illness), and ecological features (interactions of environment and host that influence disease transmission). These, and the existence of appropriate laboratory methods, must be taken into account in selecting a sampling plan.

(a) *Epidemiological considerations*

Many associations have been observed between particular foods and specific food-borne diseases. Frequently there are strong regional and cultural influences on these associations.

Water and shellfish have been shown to be common vehicles in outbreaks of typhoid fever, cholera, and hepatitis A. Meat and poultry are commonly identified as vehicles in outbreaks of salmonellosis. Ham and cream-filled pastries are frequently implicated in outbreaks of staphylococcal food-poisoning. Outbreaks of gastroenteritis caused by *Vibrio parahaemolyticus* are usually associated with seafood. Cooked meat, cooked poultry, and gravy which have undergone time-temperature abuse are the usual vehicles in outbreaks of enteritis caused by *Clostridium perfringens*. Botulism is a rare disease usually associated with the ingestion of inadequately processed home-preserved foods, particularly cured pork products, fermented fish, fish eggs, or marine mammals, and low-acid foods, including vegetables. Raw milk is commonly identified as the vehicle of campylobacteriosis, brucellosis, and salmonellosis. Additionally, brucellosis is associated with cheese made from raw milk. An association between gastroenteritis caused by *Bacillus cereus* and the consumption of cooked rice is well established.

(b) *Ecological considerations*

Many of the bacteria pathogenic to man are widely distributed in animals: for example, members of the genera *Salmonella*, *Campylobacter*, and *Yersinia* as well as *C. perfringens* and *Staphylococcus aureus*. Most food-borne disease is associated with man's consumption of animal products. Man is also a reservoir of certain food-borne pathogens, some of which may persist for weeks or months in the carrier state: for example, *Salmonella typhi*, *Shigella*, and hepatitis A virus.

Local customs and standards of community hygiene, especially those related to food, water supply, and sanitation, are important determinants of the extent and variety of food-borne diseases. Effectiveness of prevailing standards for safeguarding water supplies, milk supplies, and shellfish harvesting warrants consideration. Control of food-processing; detecting, recalling, or condemning contaminated foods; vermin control; public health supervision of food-service establishments; and intelligent use of refrigeration in processing plants, food-service establishments, and homes – all of these play a role in reducing the incidence of food-borne disease and influence the selection of sampling plans for particular commodities from particular sources.

Dietary customs peculiar to a region also influence the food-borne hazards. For example, the Japanese custom of eating fermented and raw fish has favoured the relatively high incidence of type E botulism and *V. parahaemolyticus* infection in that country.

(c) *Clinical considerations*

Certain food-borne species are inherently associated with severe illnesses in man. *Clostridium botulinum* types A, B, E, and F, for instance, can produce toxins that cause neurological illnesses when even very small amounts are ingested. If not effectively treated with antitoxins and respiratory support, the case fatality rate may exceed 50%. Virulent properties of *Salm. typhi*, *Shig. dysenteriae* I, *Vibrio cholerae*, certain strains of *Salm. typhimurium*, and *C. perfringens* type C enable these organisms to cause severe disease, even death. Cholera may present a medical emergency because in malnourished cases 50 to 70% of dehydrated cholera patients die unless they are appropriately treated by oral or intravenous fluid and electrolyte replacement.

Small doses ($10^1 - 10^2$) of *Shig. dysenteriae* and *Shig. flexneri* can cause shigellosis. The infective dose of other shigellae, *V. cholerae*, and some strains of *Salmonella* may also be low in highly susceptible individuals such as infants and malnourished and immunologically compromised persons. Also, the severity of enteritis caused by *Salmonella*, *Shigella*, and *Escherichia coli* is greater (and probably the infective dose lower) for the very young, the aged, and persons with concomitant diseases than in healthy young adults. In these groups of people, even the usually moderate gastroenteritis caused by *V. parahaemolyticus*, staphyloenterotoxin, *B. cereus*, or *C. perfringens* may sometimes become severe.

The relatively rare instances where β-haemolytic streptococci are food-borne may lead to tonsillitis complicated by severe sequelae of glomerulonephritis and arthritis, while cardiovascular disabilities (such as rheumatic fever) may follow. Convalescence after some other food-borne diseases (particularly typhoid and paratyphoid fevers, brucellosis, and viral hepatitis) may be lengthy.

(d) *Laboratory considerations*

The laboratory plays a crucial role in diagnosing botulism and enteric infections. Many physicians are unlikely to have previously encountered botulism; hence misdiagnoses may occur, even when symptoms are typical. Furthermore, effective treatment of botulism depends on early administration of appropriate antitoxin. Laboratory isolation of specific pathogens is the only way certain enteric diseases can be diagnosed because the clinical syndromes of many of these diseases are quite similar. For example, early differentiation between bacillary and amoebic dysenteries depends mainly on the laboratory.

When the existence of a particular, previously unrecognized, food-borne infection has been established by a laboratory, awareness of it increases as other incidents are revealed. The recognition of food-borne campylobacteriosis illustrates this point.

Completely satisfactory procedures are not yet available for the routine isolation and enumeration of several food-borne pathogens from foods (such as *Shigella*, *Campylobacter fetus* subsp. *jejuni*, *Y. enterocolitica*, and pathogenic *E. coli*). Non-B hepatitis viruses have not yet been isolated from foods. Laboratory methodology, therefore, influences the choice of organisms for testing and the choice of either 2-class or 3-class plans.

(e) *Conclusions*

These considerations led the Commission to conclude that food-borne pathogens should be grouped into three categories according to severity of risk. The more severe the disease, the more stringent the plan required.

Pathogens that present severe hazards (e.g., *C. botulinum* types A, B, E, and F, *Salm. typhi*, *Shig. dysenteriae* I, *V. cholerae*, *Brucella melitensis*, and *C. perfringens* type C) are grouped into a high-risk ('severe hazards') category (Table 7).

Pathogens that cause less-severe diseases have been divided into two categories. Some food-borne pathogens are of particular concern because (a) small numbers can cause disease, (b) secondary infections occur among contacts of infected persons, (c) they cause serious disease among susceptible groups of persons, (d) the initial illness can be followed by severe sequelae, or (e) they can be readily spread from raw foods of animal origin to other raw of processed foods. *Shig. flexneri*, *boydii*, and *sonnei* are pathogens that can cause disease when only small numbers are ingested and for which secondary infections occur among contacts. Pathogenic *E. coli* and certain strains or serotypes of salmonellae (e.g., *Salm. typhimurium*) can cause serious illness in the young or other highly susceptible persons such as the aged, the malnourished, or those with concurrent diseases. Severe sequelae can follow diseases caused by β-haemolytic streptococci. Salmonellae can be readily spread from infected animals to other animals and to their carcasses during processing, and subsequently from these carcasses to other carcasses, to their products after heat-processing, or to other foods in the same processing or preparation areas. All of these situations call for a category (moderate hazard, potentially extensive spread) of special concern (Table 7).

Other pathogens cause common-source outbreaks of food-borne disease, but subsequent spread is either rare or does not occur. *Bacillus cereus* and *C. perfringens*, which are almost ubiquitous, cause disease only when the ingested foods contain very large numbers of the bacteria. *Vibrio parahaemolyticus*, a frequent contaminant of marine fish of either warm sea water or colder sea waters during warm seasons, multiplies to large numbers in marine fish, shellfish, and crustacea. If staphylococcal food-poisoning occurs, *Staph. aureus* cells have at some time grown to large numbers and produced sufficient quantities of enterotoxins to elicit an emetic response. These food-borne pathogens are grouped into a less severe category (moderate hazards, limited spread; Table 7). *Campylobacter fetus* subsp. *jejuni* and *Y. enterocolitica* and *paratuberculosis* also cause common-source outbreaks without apparent secondary spread and at present seem to fit in this category. As more epidemiological and clinical data are compiled, however, these pathogens might need to be reclassified. Table 7 lists pathogens that are most frequently involved in bacterial food-borne diseases, and is not intended to be exhaustive. No attempt has been made to arrange these pathogens according to the frequency with which they cause outbreaks or cases of food-borne disease, since this varies with locality.

Also omitted from the table are agents that cause viral and parasitic diseases, mycotoxicoses, poisons of certain plants or fish and other animals, and chemical toxicants. These problems are outside the scope of this book. For information on these agents see Bryan, 1982; Riemann

TABLE 7

Major food-borne pathogens or toxins[a]

Organism (disease)	Frequency of involvement in disease	Examples of vehicles	Other factors contributing to significance
I. SEVERE HAZARDS			
Brucella melitensis, B. abortus, B. suis (brucellosis)	Moderately rare, but occasionally endemic	Raw milk and cheese, particularly from goats	Convalescence often prolonged
Clostridium botulinum types A, B, E, and F (botulism)	Rare	Improperly processed canned or preserved low-acid foods; 'home' cured meat products; smoked fish, fermented fish (eggs), and other marine products	High mortality; rapid recognition and treatment may be essential for patient survival; precautions needed by processors
Clostridium perfringens type C (*Enteritis necroticans*)	Rare	Cooked pork	High mortality of protein-deficient persons
Hepatitis A virus (hepatitis A)	Common	Raw shellfish, salads, cold cuts of meat, raw vegetables, water	Very severe for patients with liver disease; convalescence prolonged
Salmonella typhi, S. paratyphi A, B (*schottmulleri*), and C (typhoid and paratyphoid fevers)	Endemic in many parts of the world; occasionally epidemic	Water, raw milk and milk products, meat products, raw vegetables, shellfish	Prolonged medical care required; liable to result in carrier state (especially *S. typhi*)
Shigella dysenteriae I (shigellosis, bacillary dysentery)	Sporadic or epidemic	Raw vegetables, salads	High mortality rate; frequently misdiagnosed; low infective dose
Vibrio cholerae O1, (*cholera*)	Sporadic, endemic, and occasionally epidemic	Fish, shellfish, water	

TABLE 7 (continued)

Major food-borne pathogens or toxins[a]

Organism (disease)	Frequency of involvement in disease	Examples of vehicles	Other factors contributing to significance
II. MODERATE HAZARDS: POTENTIALLY EXTENSIVE SPREAD[b]			
β-haemolytic *Streptococcus* group A, C, G (septic sore throat, scarlet fever)	Infrequently food-borne	Raw milk and milk products, salads containing eggs	Higher incidence in malnourished and those recovering from intestinal surgery; high mortality of dehydrated, untreated persons Sequelae of actue pharyngitis, nephritis, arthritis, and cardiovascular complications; secondary spread between contacts
Pathogenic *Escherichia coli* (*E. coli* diarrhoea)	Increasingly reported from many areas; important cause of travellers' diarrhoea	Foods subjected to contact *with infected persons*; raw or rare meats, raw milk and milk products, salads, water	Serious for the young; shigella-like syndrome
Salmonella typhimurium and other *Salmonella* serovars (salmonellosis)	Very common	Poultry, eggs, meats, wide range of other foods	Serious for young and old persons; cross-contamination from raw meat and poultry; some strains of *S. typhimurium* and other serovars highly virulent
Shigella flexneri S. boydii, S. sonnei (see Section I) (shigellosis, bacillary dysentery)	Sporadic, sometimes endemic	Foods subjected to contamination by infected persons or sewage-contaminated water such as raw vegetables, salads	Serious for young and old persons; secondary infections among contacts; sometimes low infective doses

TABLE 7 (concluded)

Major food-borne pathogens or toxins[a]

Organism (disease)	Frequency of involvement in disease	Examples of vehicles	Other factors contributing to significance
III. MODERATE HAZARDS: LIMITED SPREAD[b]			
Bacillus cereus (*B. cereus* gastro-enteritis)	Common	Fried and boiled rice, reconstituted cereal products, puddings and custards	
Campylobacter fetus subsp. *jejuni*[c] (campylobacteriosis)	Common	Raw milk, water, poultry, meat	
Clostridium perfringens type A (*C. perfringens* enteritis)	Common	Cooked non-cured meats, poultry, and gravy	
Staphylococcus aureus (staphylococcal enterotoxicosis or food-poisoning)	Common	Ham, fermented sausages; other meat and poultry products; cream-filled pastries; raw milk, cheese and egg dishes; casein confectionery	
Vibrio cholera, non-O1 (non-O1 cholera)	Increasingly reported	Shellfish, crustacea	
Vibrio parahaemolyticus (*V. parahaemolyticus* gastroenteritis)	Common in Asia; increasingly reported elsewhere	Raw marine fish; crustacea	
Yersinia enterocolitica and *Y. paratuberculosis* (yersiniosis)	Increasingly reported	Raw milk, recontaminated pasteurized milk, water, tofu	Symptoms may simulate appendicitis

a For fuller details of food-borne diseases, consult Riemann and Bryan, 1979; Bryan, 1982.
b See text (Conclusions) for explanation of extensive and limited spread.
c *Bergey's Manual*, 8th edition; divided into two species in Véron and Chatelain (1973) and Approved Lists of Bacterial Names of the International Commission on Systematic Bacteriology (Skerman et al., 1980).

and Bryan, 1979; Halstead, 1978; NAS-NRC, 1973; ICMSF, 1978, pp. 68–89.

C. AN INDEX OF UTILITY

Utility tests – for general contamination, shelf-life, or spoilage, for example – are not related to health hazard but rather to economic and aesthetic considerations. Therefore, the corresponding cases 1–3 in Table 6 (p. 43) (cf. Table 9, p. 58) can be satisfied by relatively lenient plans. They may involve direct microscopic counts, yeast and mould counts, aerobic plate counts, or specialized tests such as for cold-tolerant organisms or for species recognized as causing a particular type of spoilage (e.g., lactobacilli in mayonnaise, or thermophilic spore-formers in sugar).

D. INDICATOR TESTS (low hazard)

An inspector at a port of entry usually knows very little about the history of a consignment of food. He does not know whether a heat process adequate to kill relevant microorganisms was used in processing, whether the food has been contaminated after processing, or whether the consignment has been temperature-abused during shipment. Tests for relevant organisms can indicate the history of processing hygiene for certain foods. For instance, high numbers of mesophilic spore-forming bacteria in non-acid, shelf-stable canned foods indicate probable underprocessing. Sufficient numbers of Enterobacteriaceae or coliforms can indicate inadequacy of general hygiene in a food plant. *E. coli* suggests faecal contamination, and staphylococci contamination from the human skin or nose (see also ICMSF, 1978). In some instances, indicators such as *E. coli* may be used to point to potentially severe hazard (e.g., in shellfish-growing waters). Because of the uncertainty of relationships between indicators and specific pathogens, it is considered inappropriate to apply more stringent cases for indicator organisms at this time.

E. TESTS FOR PATHOGENS (moderate and severe hazard)

Tests for particular pathogens should be applied (a) in routine sampling when the history of sample analysis shows that when the pathogen is present it causes illness, and testing is an effective means of consumer protection; (b) in investigational sampling: (i) when the epidemiology of a food-borne disease outbreak points to a particular 'lot' of food as the cause of illness, and (ii) when there are other circumstances creating suspicion of the presence of the pathogen.

The cases for pathogens are further classified in Table 6 (p. 43) into 'moderate hazard' and 'severe hazard.' Table 7 (p. 48) provides more information regarding these hazardous organisms.

In special circumstances, moderately hazardous organisms may constitute a severe hazard. For example, as shown in Table 8 (p. 55), very young, very old, or debilitated persons may become seriously ill or even die from salmonellosis; and persons on salt-free diets may become severely ill through loss of blood potassium simply because of diarrhoea.

The choice of plan to suit the difficult cases where special hazards exist is considered in Chapter 6; further illustration of factors influencing hazard in relation to sampling plans is given in Section B of that chapter.

F. CHOOSING APPROPRIATE TESTS

In choosing appropriate tests for particular foods, one needs to consider the following:

1 The food-borne disease record of the type of food. For example, egg products have a history of causing *Salmonella* infections but not botulism. Thus the plans for egg products include a test for *Salmonella* but not for *C. botulinum*.
2 The probable past history of processing and storage. For example, it is appropriate to test refrigerated cooked beef for *Salmonella* but not shelf-stable uncured canned beef.
3 The expectation of subsequent spoilage. For example, tests are commonly conducted to predict the keeping quality of fluid milk.
4 The age of the lot. For example, different tests are relevant to natural mineral water at the point of bottling and in distribution.

Such considerations will indicate the species of microorganisms likely to present hazards in a particular food. The choice of test methods, suitable for the various species likely to be involved, is discussed in ICMSF (1978). The preceding sections of this chapter indicate the associated type of hazard, and the corresponding horizontal row in Table 6 indicates the case to be applied.

G. THE EFFECT OF CONDITIONS OF USE ON DEGREE OF HAZARD

As noted in Section A of this chapter, hazard is related to the numbers of an undesirable organism present in a food. To choose an appropriate hazard column in Table 6 (p. 43), the effect of the conditions to which the food will probably be exposed on the growth or death of the relevant microbes in the food must be considered.

If the food is ordinarily subjected to conditions that permit bacterial growth, thereby increasing hazard, the case would be 3, 6, 9, 12, or 15. For example, one of these would apply to dried milk since contaminating pathogens may multiply after reconstitution.

If conditions would not cause a change in the number of relevant bacteria (e.g., frozen storage), the appropriate case would be 2, 5, 8, 11, or 14. Ices (e.g., ice cream) would be classified in one of these cases, because they are ordinarily consumed frozen.

If a food is expected to be fully cooked later, one would choose case 1, 4, 7, 10, or 13 since cooking reduces the hazard. Poultry, pasta, and many dried products fall into this category.

Other factors to be considered include conditions of further storage of the food (i.e., frozen, at ambient temperatures, or in a hot warehouse), further processing (i.e., cooking, fermenting, washing, or freezing), and consumption habits (i.e., normally eaten raw, warmed, or cooked). Such consideration can take into account only the usual treatments the lot is expected to endure. For example, a frozen food will ordinarily be kept frozen until thawed for cooking or serving. If a food is unexpectedly abused after being sampled and approved, the conditions envisaged will, of course, not apply.

For a particular lot of food, there will usually be one most appropriate column in Table 6, but occasionally different columns will apply to the different microbial species present according to their ability to grow, survive, or die in the food.

In deciding whether conditions of use will change the hazard, the expected treatment must be considered in relation to the nature of the food in question, as it will affect each relevant bacterial species. The following examples illustrate such considerations.

First, the nature of the food should be considered in relation to the growth requirements of the relevant species. Many foods with a salt concentration of c. 10% may support the growth of staphylococci but not that of salmonellae. Thus, such products (if not refrigerated) might be classified in case 9 with respect to staphylococci and in case 11 for salmonellae.

Fresh meat supports the growth of various bacteria, whereas dried meat does not; hence, if stored at ambient temperatures, hazard would increase in fresh meat corresponding to case 3, 6, 9, 12, or 15, whereas for dried beef there would be no change in hazard from bacteria and case 2, 5, 8, 11, or 14 would apply.

Growth of an organism of concern may sometimes be prevented by competition from other microorganisms. While salmonellae grow in most foods of appropriate pH and a_w, staphylococci are often restricted by associated spoilage bacteria. Raw meats and bacon are not normally associated with staphylococcal food-poisoning because they also carry large numbers of competing organisms. The hazard of enterotoxin formation usually arises in cooked foods, which generally contain few microorganisms other than staphylococci (e.g., cooked ham contaminated after cooking, or cheese and fermented meats where the competitive fermentative activity is delayed).

Temperature relations are especially important. Microbial numbers, and associated hazards, generally increase at ambient temperatures and increase even more at warmer temperatures. In contrast, refrigeration temperatures tend to control the hazard, since many pathogens do not multiply at low temperatures: for example, for ham kept below 6°C (at which temperature staphylococci do not produce toxin) case 8 rather than case 9 would apply. The nature of the food and its associated species may be significant. For example, storage of chilled smoked fish for long periods could permit toxin production by *C. botulinum* type E, because this organism can grow at about 3°C. Prolonged refrigeration of canned perishable ham creates no known hazard, however.

Eating customs also affect hazard and the choice of 'case.' For example, *V. parahaemolyticus* grows readily on raw fish unless it is refrigerated. It is a major cause of food-poisoning in Japan, where the eating of raw fish is common; but in North America, though widely distributed, it is a much less common cause of illness. Thus, for Japan case 8 or 9 would be appropriate for this species, whereas in North America case 7 might be. Note that foods that are cooked before distribution (e.g., powdered eggs) are sometimes eaten without cooking when distributed in relief foods. If a food is intended for consumers with unusually high susceptibility to food-borne disease, the hazard will be increased (see Table 8).

When choosing cases on the basis of hazard as above, one must consider whether the consignment under examination will be used

TABLE 8

Special foods for consumer groups with increased suseptibility

Food class	Reason for stringent sampling plan
Baby foods	High susceptibility of the consumer population to enteric pathogens; severe response to infections and toxins; increased risk of mortality
Dietetic foods	Infection is a severe risk for diabetics.
Foods for hospitals	(a) Patients may be prone to infection and to serious sequelae after enteric disease because of stresses from other disabilities and from immunosuppressive treatment, and under intensive care. (b) Interference with convalescence from other disease. (c) Staff and patients need to be protected because of their potential for spreading the disease within the hospital.
Relief foods, especially dehydrated high-protein foods	Populations needing relief foods are usually highly susceptible and prone to serious complications because of malnutrition and other stressful conditions. There is also increased risk for person-to-person spread of disease owing to confinement of the population in crowded areas often having poor sanitary conditions. Particular hazards are reconstitution with contaminated water, unhygienic handling, and poor storage conditions leading to rapid bacterial growth.

exclusively for a single purpose. If a consignment might include lots to be used for a variety of purposes, it should meet the criterion consistent with the general use of the product. Exception can be made if a specific use of the product is known and a more stringent criterion is needed.

H. CHOOSING THE 'CASE'

In general, foods that have received heat treatment are safer than those that have not. Hazard is likely when foods (i) become contaminated after processing, (ii) are exposed to conditions that permit multiplication of pathogens, and (iii) are not cooked shortly before consumption. These points must always be considered when estimating the degree of hazard in choosing the case and sampling plan.

Foods may be used in diverse ways. For example, dried egg may be cooked and eaten at once (e.g., scrambled egg, custard) or cooked as a component of products that are eaten later but are not subject to multiplication of pathogens (e.g., cake, pasta); both these treatments diminish microbiological hazards. It may be eaten dry (e.g., dusting powder), or reconstituted and eaten promptly; these treatments do not change the hazard. Sometimes, there may be long delays between reconstitution and consumption, and the hazard increases.

With an unpasteurized product an aerobic plate count (APC) would provide a useful index, but with a pasteurized product APC reflects the result of the process so that an offensively spoiled raw material might meet normal APC criteria after pasteurization. A direct microscopic count (DMC) is a desirable additional test for the number of bacteria present before pasteurization. With DMC or APC, the choice of case (1, 2, or 3) depends simply on subsequent use.

The choice of case for the indicator organisms – coliforms, faecal coliforms, Enterobacteriaceae, or *E. coli* – is aided by the following considerations. Raw foods may contain various species of these organisms in substantial numbers. Their presence indicates general contamination, but some tolerance is necessary because, given the nature of the materials and the operations, it may not be possible to produce the food without coliforms. If a criterion is to be used in foods which are subsequently heat-processed, Case 4 is appropriate for the raw food. Differentiation within the coliform group is unnecessary because all gram-negative bacteria will be destroyed during thorough cooking, and the general level of contamination is the only concern.

In contrast, when a food has been pasteurized, coliforms will have been destroyed. Any that are present will usually be derived from post-processing contamination, and indicate poor and unacceptable hygiene. If a product (e.g., reconstituted dried milk or eggs) is subjected to delay after reconstitution, case 6 is appropriate. Therefore, selection of *E. coli* or faecal coliforms may be desirable, the case (4, 5, or 6) being determined by expected conditions of subsequent use.

With the moderate hazard provided by coagulase-positive staphylococci (i.e., probably enterotoxin-forming), one could be lenient towards raw products that are to be cooked (case 7). Competition from the natural microflora will limit multiplication of staphylococci, making the production of enterotoxin unlikely. However, foods that have been cooked are frequently contaminated with staphylococci by human handlers. As the original flora has been largely inactivated, the staphylococci experience little competition and, if conditions are favourable, growth and toxin production may ensue. Case 9 is then appropriate (e.g., for ready-to-eat frozen bakery products, Table 29; or cooked chilled or frozen crabmeat, Table 27). Pasteurized foods to be eaten frozen (e.g., ice cream) or immediately after thawing would be allotted to case 8. In certain salted foods in which salt-tolerant *Staph. aureus* can grow, the competing flora is inhibited by the reduced a_w. The cases for staphylococci will be similar to those suggested in pasteurized foods.

Bacillus cereus and *C. perfringens* are also moderate hazards, differing importantly from *Staph. aureus* in that they produce spores which survive mild heating. Few processes will reduce the hazard provided by these bacteria, so case 8 or 9 is usually appropriate, depending on whether the food is eaten immediately after either thawing or reconstitution or there is a delay between these preparation steps and eating.

The moderately to seriously hazardous salmonellae often occur in raw protein foods (e.g., egg pulp) but should be destroyed by pasteurization. However, recontamination of pasteurized products with salmonellae is not uncommon, and subsequent drying or freezing cannot be relied upon to destroy these bacteria. Consequently it is frequently advisable to test for *Salmonella* in this type of product. If such a dried food is consumed dry, there is no change in hazard (case 11); if reconstitution is delayed, and heating does not take place before consumption (a practice that is highly undesirable with many such products), the case would be 12. Cooking promptly after reconstitution will reduce the hazard, so case 10 would then be appropriate.

The foregoing illustrates and emphasizes the need for some knowledge of the microbial ecology of a food before an examiner can choose an appropriate case and test for a particular purpose. Applying the principles outlined, Table 9 gives numerous examples of the relations between cases and conditions of handling and usage, and the organisms of concern, for various foods illustrating the cases considered in Part II.

I. DECIDING BETWEEN TWO-CLASS AND THREE-CLASS PLANS

To decide whether the plan should be 2-class or 3-class, one must consider whether any positives (e.g., *Salmonella* or APC levels above those reflected by good manufacturing practice) can be permitted in any of the sample units. If the answer is no, a 2-class plan with $c = 0$ should be used. If the answer is yes, a 2- or 3-class plan can

TABLE 9

Relation of 'cases' (degree of concern)[a] to food type, microbiological test, and conditions of food treatment – illustrative examples from various classes of foods

Nature of concern	Effect of conditions	Case	Food product	Bacterial test	Anticipated conditions of treatment
No direct health hazard, general contamination, spoilage, and shelf-life	Reduce hazard	1	Fish, fresh or frozen	APC	Cooked before eating
		1	Raw crustaceans, frozen	APC	Cooked before eating
		1	Carcass meat before chilling[b]	APC	Cooked before eating
		1	Carcass meat, chilled[b]	APC	Cooked before eating
		1	Edible offals, chilled[b]	APC	Cooked before eating
		1	Carcass meat, frozen[b]	APC	Cooked before eating
		1	Boneless meat, frozen[b]	APC	Cooked before eating
		1	Comminuted meat, frozen[b]	APC	Cooked before eating
		1	Edible offal, frozen[b]	APC	Cooked before eating
		1	Raw chicken, fresh or frozen[b]	APC	Cooked before eating
	No change in hazard	2	Dried milk	APC	Consumed or refrigerated immediately after rehydration
		2	Pasteurized liquid, frozen, or dried egg products	APC	Used immediately upon reconstituting in formulated products which are not further heat-treated
		2	Precooked, breaded fish products, frozen	APC	Eaten promptly upon thawing
		2	Cooked crustaceans, frozen	APC	Eaten promptly upon thawing
	Increase hazard	3	Cooked crabmeat, chilled or frozen	APC	Recooked before eating, though perhaps subjected to poor temperature control and held for extended periods of time after opening or thawing

		3	Bivalve molluscs, fresh or frozen	APC	Eaten raw or after light cooking only, though perhaps subjected to poor temperature control and held for extended periods after opening or thawing
Low indirect health hazard (indicator organisms)	Reduce hazard	4	Fish, fresh or frozen	*E. coli*	Cooked before eating
		4	Raw crustaceans, frozen	*E. coli*	Cooked before eating
		4	Dried products that require heating to boiling before eating	APC, coliforms	Cooked before eating
	No change in hazard	5	Dried milk	Coliforms	Consumed promptly upon rehydration
		5	Pasteurized liquid, frozen, or dried egg products	Coliforms	Used immediately without further processing
		5	Precooked, breaded fish products, frozen	*E. coli*	Not always thoroughly reheated before eating, but eaten promptly upon reheating
		5	Cooked crustaceans, frozen	*E. coli*	Eaten promptly upon thawing
		5	Vegetables or fruits (pH 4.5), frozen	*E. coli*	Eaten promptly upon thawing
		5	Dried vegetables	*E. coli*	Eaten promptly upon rehydration
		5	Cereals (including flours)	Moulds	Cooked but mycotoxins not necessarily destroyed by heating
		5	Soya flour, concentrates, and isolates	Moulds	Further processed, but mycotoxins not necessarily destroyed by heating
		5	Coated or filled, dried shelf-stable biscuit	Coliforms	Eaten in the dry state
		5	Non-carbonated natural mineral water (*eaux plates*) and other bottled non-carbonated waters	Coliforms	Usually consumed promptly upon opening (organisms decrease in time in bottle, but some may survive)

TABLE 9 (continued)

Nature of concern	Effect of conditions	Case	Food product	Bacterial test	Anticipated conditions of treatment
Low indirect health hazard (indicator organisms)	Increase hazard	6	Cooked crustaceans, frozen	*E. coli*	Little or no cooking before eating; sometimes stored under conditions of poor temperature control and held for an extended period of time after thawing
		6	Cooked crabmeat, chilled or frozen	*E. coli*	Little or no cooking before eating; sometimes stored under conditions of poor temperature control and held for an extended period of time after thawing
		6	Bivalve molluscs, fresh or frozen	*E. coli*	Little or no cooking before eating; sometimes stored under conditions of poor temperature control and held for an extended period of time after thawing
		6	Dried and instant formulated products that require reconstitution	APC,[c] coliforms	Perhaps subjected to poor temperature control after reconstitution and held for extended periods of time
Moderate direct health hazard, limited spread	Reduce hazard	7	(No example in current ICMSF criteria)	Vegetative pathogen that does not produce heat-stable toxin	Cooked before eating
	No change in hazard	8	Dried blood, plasma, or gelatin	*Staph. aureus*, *Cl. perfringens*	Used as an ingredient in products not receiving adequate heat-treatment (enterotoxin and spores survive heating)
		8	Cooked poultry meat, to be cooked, frozen	*Staph. aureus*	Reheated before eating, but enterotoxin would survive
		8	Cooked poultry meat, ready-to-eat (e.g., turkey rolls), frozen	*Staph. aureus*	Eaten promptly after thawing or held cold

	8	Cooked crustaceans, frozen	*Staph. aureus*	Eaten promptly upon thawing
	8	Cheese, 'hard' and 'semi-soft' types	*Staph. aureus*	Not normally heated, may not be refrigerated (no growth should develop after processing)
	8	Bakery products containing low-acid or high a_w fillings (e.g., meat pies or toppings, pizzas), frozen	*Staph. aureus*	Not always thoroughly reheated before eating, but promptly eaten after reheating
	8	Entrées containing rice or corn flour as a main ingredient	*B. cereus*	Reheated, but spore and emetic toxin survive heating; eaten promptly after reheating
	8	Pasta, frozen or dried	*Staph. aureus*	Normally heated, but enterotoxin would survive
	8	Non-carbonated natural mineral water (*eaux plates*)	*Pseudomonas aeruginosa*	Normally consumed promptly after opening; no further treatment
Increase hazard	9	Cooked poultry meat, ready-to-eat (e.g., turkey rolls), frozen	*Staph. aureus*	Not always reheated; poor temperature control after thawing and held for extended periods of time
	9	Cured and/or smoked poultry meat	*Staph. aureus*	Perhaps subjected to poor temperature control and held for extended periods of time
	9	Dried and instant products that require reconstitution	*Staph. aureus*	Perhaps subjected to poor temperature control and held for extended periods of time after reconstitution
	9	Bakery products (ready-to-eat) with low-acid or high-a_w ingredients, frozen	*Staph. aureus*	Perhaps subjected to poor temperature control and held for extended periods of time after thawing
	9	Cooked crabmeat, chilled or frozen	*Staph. aureus*	Perhaps not cooked before eating and subjected to poor temperature control and held for extended periods of time after thawing

TABLE 9 (continued)

Nature of concern	Effect of conditions	Case	Food product	Bacterial test	Anticipated conditions of treatment
Moderate direct health hazard, of potentially extensive spread in food	Reduce hazard	10	Dried milk	*Salmonella*	Used as an ingredient of a heat-processed product
		10	Pasteurized liquid, frozen, or dried egg products	*Salmonella*	Used as an ingredient of a heat-processed product
		10	Dried products that require heating to boiling before comsumption	*Salmonella*	Heated before eating
		10	Cooked poultry meat to be reheated before eating (e.g., prepared meals), frozen	*Salmonella*	Reheated before eating
		10	Soya flours, concentrates, and isolates	*Salmonella*	Used as an ingredient in a heat-processed product
		10	Bakery products (to be cooked) with low-acid or high-a_w fillings (e.g., meat pies or toppings, pizzas), frozen	*Salmonella*	Heated before eating
		10	Cereal products, frozen or dried	*Salmonella*	Heated before eating
	No change in hazard	11	Cooked poultry meat, ready-to-eat (e.g., turkey rolls), frozen	*Salmonella*	Eaten promptly after thawing
		11	Cured and/or smoked poultry meat, refrigerated or frozen	*Salmonella*	Eaten promptly after opening or thawing
		11	Dehydrated poultry products	*Salmonella*	Perhaps not thoroughly reheated or used in salads after rehydrating; eaten promptly after rehydrating
		11	Dried milk	*Salmonella*	Eaten promptly after rehydrating
		11	Pasteurized liquid, frozen, or dried egg products	*Salmonella*	Eaten promptly after rehydrating or thawing
		11	Cocoa	*Salmonella*	Used as an ingredient in a shelf-stable product

	11	Chocolate and other confectionery	*Salmonella*	Shelf-stable product eaten without further treatment
	11	Intermediate-moisture pet foods	*Salmonella*	Shelf-stable product fed without further treatment
	11	Dry pet foods, not-to-be reconstituted	*Salmonella*	Shelf-stable product fed without reconstituting
	11	Dried blood, plasma, or gelatin	*Salmonella*	Used in a shelf-stable product
	11	Coconut (desiccated)	*Salmonella*	Eaten in the dry state; growth not anticipated
	11	Peanut butter	*Salmonella*	Shelf-stable product eaten without further treatment
Increase hazard	12	Pâté	*Salmonella*	Perhaps subjected to poor temperature control and held for an extended period of time
	12	'Roast' beef, frozen	*Salmonella*	Perhaps subjected to poor temperature control and held for an extended period of time after thawing
	12	Dehydrated pet food to be reconstituted	*Salmonella*	Perhaps subjected to poor temperature control and held for an extended period of time after rehydrating
	12	Dried milk	*Salmonella*	Perhaps subjected to poor temperature control and held for an extended period of time after rehydrating
	12	Pasteurized liquid, frozen, or dried egg products	*Salmonella*	Perhaps subjected to poor temperature control and held for an extended period of time after thawing or rehydrating

TABLE 9 (continued)

Nature of concern	Effect of conditions	Case	Food product	Bacterial test	Anticipated conditions of treatment
Moderate direct health hazard, of potentially extensive spread in food	Increase hazard	12	Coconut (desiccated)	*Salmonella*	Likely used in a high-moisture product; poor temperature control and held for an extended period of time after formulating
		12	Yeast	*Salmonella*	Perhaps used as an ingredient in a high-moisture product; poor temperature control and held for an extended period of time after formulating, thawing, or rehydrating
		12	Peanut butter	*Salmonella*	Perhaps used as an ingredient in a high-moisture product; poor temperature control and held for an extended period of time after formulating in a high-moisture product
		12	Dried and instant formulated products that require reconstitution	*Salmonella*	Perhaps subjected to poor temperature control and held for an extended period of time after reconstituting
		12	Bakery products (ready-to-eat) with low-acid or high a_w fillings or toppings, frozen	*Salmonella*	Perhaps subjected to poor temperature control and held for an extended period of time after thawing

Severe direct health hazard	Reduce hazard	13	(No example in current ICMSF criteria)	*Brucella*, *C. botulinum*, *C. perfringens* type C, *Salm. typhi*, *Shig. dysenteriae*, *V. cholerae*, hepatitis A virus	Cooked before eating
	No change in hazard	14	(No example in current ICMSF criteria)	*Brucella*, *C. botulinum*, *C. perfringens* type C, *Salm. typhi*, *Shig. dysenteriae*, *V. cholerae*, hepatitis A virus	Shelf-stable product eaten in processed state or promptly after thawing or rehydration
	Increase hazard	15	(No example in current ICMSF criteria)	*Brucella*, *C. botulinum*, *C. perfringens* type C, *Salm. typhi*, *Shig. dysenteriae*, *V. cholerae*, hepatitis A virus	Subjected to poor temperature control and held for an extended period of time after preparing, thawing, or rehydrating

a Compare with Table 6, p. 43.
b For in-plant quality control; not for use at port of entry.
c APC considered as an indicator of health risk.

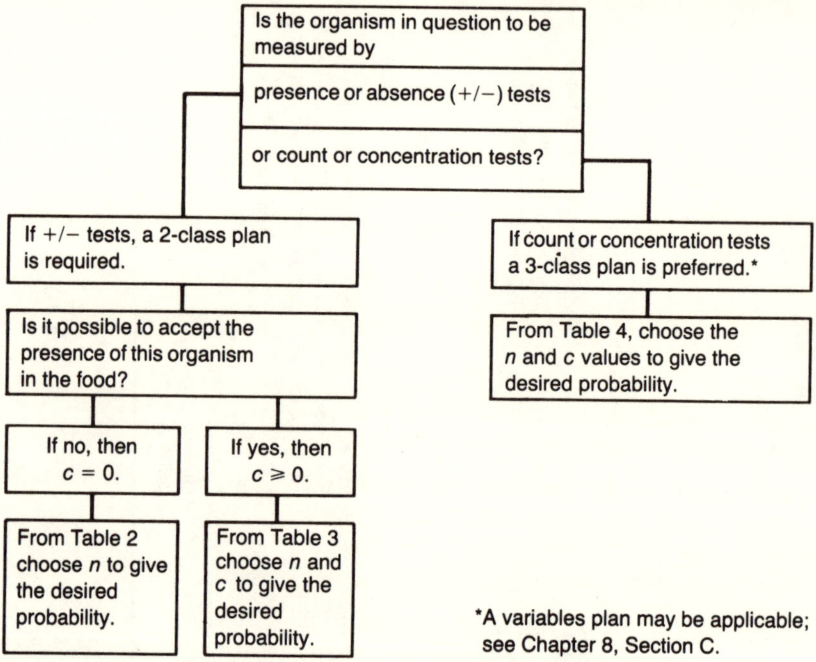

Figure 3 Choosing sampling plan for a specific application

apply, but if the number of microbes in a unit-volume or mass can be obtained a 3-class plan is recommended (Fig. 3).

Three-class sampling plans, where applicable, have properties that should encourage their adoption in microbiological examination, for the following reasons:

(a) To accept a proportion of sample units yielding test values in the marginally acceptable interval (between acceptable and defective), as these plans do, is in keeping with practical experience where, even under good manufacturing conditions, a few sample units may well reveal test values beyond those normally encountered without causing any consequent problem. This situation applies especially to counts of 'indicator' organisms.

(b) Sufficient experience also defines a different level, beyond which counts indicate substantial likelihood of health or utility hazard, a level which will not be attained if control has been adequate. This, designated M, should remain stable unless new experience reveals an error in its initial placement. Obviously, greater stability of criteria of acceptance may be expected to promote wider adoption of sampling plans.

(c) A 3-class plan allows a regulatory authority to carry out a limited form of trend analysis. For example, an increasing proportion of values within the marginally acceptable (*m* to *M*) range may indicate a lack of control.

J. DETERMINING VALUES FOR *m* AND *M*

Preliminary definitions of *m* and *M* have been given in Chapter 4.

The level of the test organism which is acceptable and attainable in the food is *m*. It is reflected by Good Commercial Practice (GCP) as recognized either in the exporting country or, more often, in the domestic production of the importing country, or is the hazardous level in 2-class plans. If the test organism is a pathogen (i.e., for 2-class plans) *m* may be zero or, more correctly, a small number corresponding to the level of detectability in the test (see Chapter 6, Section B). Hence, in this book, the value of *m* for 2-class plans is usually 0. For 3-class plans *m* will usually be assigned some non-zero value.

M, used only for 3-class plans, is a hazardous or unacceptable level of contamination caused by poor hygienic practice, including improper storage. All lots with values exceeding *M* should be rejected (i.e., in the sense defined in Section H of Chapter 2, they should be withheld pending further investigation to determine whether the product can be used after further treatment or is unacceptable as food). When such lots are detected, investigation of the producer's facilities should be undertaken immediately. In international commerce this presupposes a system whereby the analytical results can be conveyed to appropriate officials of the country of origin, though at present systems are largely ad hoc or non-existent.

There are several approaches to choosing the value of *M*:

1 As a utility (spoilage or shelf-life) index. Relate levels of bacteria to detectable spoilage (odour, flavour) or to a decrease in shelf-life to an unacceptably short period.
2 As a general hygiene indicator. Relate levels of the indicator bacteria to a clearly unacceptable condition of hygiene – whether contamination or growth, or a combination of these factors.
3 As a health hazard. Relate levels of bacteria to illness. Use epidemiological and laboratory data in combination, experimental an-

imal feeding or inoculation data, human feeding data, laboratory analyses for toxin related to levels of organisms, or other guides that show the level at which there is a serious concern for the hazard. For this purpose, consider the maximum amount of food likely to be eaten at one time and the sensitivity of the group of persons likely to eat the food (see also Table 8).

It is important to realize that M is defined by the hazard. The value of m is available and defined by GCP, and may alter with time. The values m and M are independent of each other and have no constant relationship. In the past, 3-class plans have been applied incorrectly by fixing the relationship between m and M.

All probability computations (e.g., OC curves and surfaces) for sampling plans assume that the laboratory result is obtained without error. The decision to accept or reject any particular lot will, of course, be based upon the particular laboratory results obtained, and is subject to the error associated with the laboratory procedure used. It is especially important when using 2-class plans with $m = 0$ and $c = 0$ that the methods employed give accurate answers, for whatever false negatives and false positives occur will result in corresponding wrong decisions about lots because of the inaccurate methods.

Unfortunately, documentation of the inherent variation within a given method is rare, though experienced analytical bacteriologists develop estimates which become included in their choice of criteria. Some methods, for example MPN methods for coliforms, have wide inherent variability (see Silliker et al., 1979). Similarly, it is commonly found that long familiarity of an analyst with a particular method (e.g., for *Salmonella* or *Staph. aureus*) allows that method to be most productive for him. This is a cogent argument for widely accepted use of a standard method so that all analysts become experienced in its use: another method with different sensitivity might require change of criteria. Because the testing method is so important in deciding the acceptance criteria of sampling plans, it seems desirable not to change a standard method unless very real advantage in so doing has been demonstrated.

Figure 4 illustrates the effect of various frequency distributions of microbial content, within a lot, on the location of m and M for 3-class plans. Curve 1 represents an entirely satisfactory lot, with small numbers of bacteria generally, and thus a low average count with little variation, so that no counts exceed m. Curve 2 represents a lot with

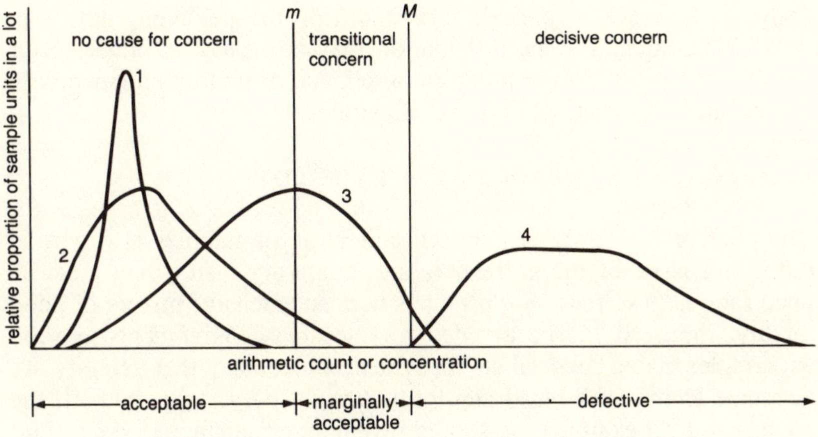

Figure 4 Considerations in choosing m and M

a similar average count, but with a much wider variation, so that a small proportion of sample units would have counts exceeding m, though none exceed M. If the proportion in the range m to M were small, the situation would be acceptable; if this proportion were larger it might still be acceptable, but it would call for a warning to the producer, as tending toward the situation shown in curve 3. Curve 3 represents a lot with a higher average count and large variation, such that a small proportion of sample units exceeds M and would cause immediate rejection, while a substantial proportion falls in the range m to M, which itself could also suffice to justify rejection (see Section H, Chapter 2, for explanation of the term 'rejection'). Curve 4 represents a lot of even greater unacceptability, requiring rejection. In an unusual situation one might find a consignment composed of two widely dissimilar lots, one entirely acceptable (curve 1) and the other entirely unacceptable (curve 4); this situation would show a high proportion of units falling below m and above M, but few between m and M. These examples illustrate how the choice of m and M, in relation to the microbial quality of the lot, influences the proportions of material likely to be rejected. Moreover they illustrate the importance to producers of limiting variation in numbers of microbes present in food, that is, of keeping production 'under control' in the bacteriological sense, to avoid lots of otherwise satisfactory average quality being rejected. With good or bad control, the average microbial levels are likely to be respectively low or high, and the distribution of the data above the average (expressed logarithmically) respectively narrow or broad. The spread of the distribution may be nearly as important as the average level in determining compliance or non-compliance with a particular criterion, as Figure 4 illustrates.

K. SPECIFIC KNOWLEDGE ABOUT THE LOT

The relatively long delay between collection of samples at a port of entry and issue of the corresponding laboratory result may create a need for costly storage. For products with an extensive history of good quality, the need can be avoided by releasing consignments as soon as samples have been taken (provided, of course, that records are adequate to trace for recall any lot that might prove to be bad). If the bacteriological examination should reveal an unsatisfactory condition, future consignments would be held at the port until a sequence of tests (on, say, three consecutive consignments) proved satisfactory. For products receiving only indicator tests, examination might be waived except for, say, a fifth of the shipments, chosen at random. These systems have been in use by various countries.

When a food product is produced and shipped under uniform and adequate controls, a comparatively small routine sample (e.g., n = 5) can be used. Until such assurance has been obtained through good past history and/or surveillance of some sort, the only way to obtain satisfactory protection (i.e., substantial discrimination between good lots and bad lots) is by increasing the sample size (larger n). But, as noted earlier (Chapter 3, Section E(e)), to double the reliability may require four times as many sample units. The cost of the additional testing should be balanced against the potential gain in discriminating power, and a practical decision made.

Food commonly is shipped in carloads or large cases containing smaller packages or units. In such cases, there is need to sample as widely as possible and at random among the carloads or cases, and then to sample randomly a few of the packages among those chosen. This method is preferable to that of choosing a few carloads or cases and many packages from each chosen carload or case.

It may not be feasible to sample at random over the entire consignment. It may, in fact, only be possible to sample randomly from a portion of the consignment called the 'frame' (see Chapter 3, Section D). If so, then results apply only to the frame, not to the consignment, and it is for the regulatory agent to judge how far the results of the frame can be trusted to apply to the entire consignment. For example, accessible containers may be those nearest the door of a vehicle, nearest the hatch of a ship's hold, or on the periphery of a stack in a warehouse. The sample units chosen from such frames may sometimes represent the parts of a lot or consignment which have been exposed to greater hazard, either of contamination or of multiplication of existing organisms, in which case the use of the 'frame' samples would provide greater consumer protection through the selection of those most likely to be hazardous. One thing working in favour of reliable results is that there often is unconscious or unplanned randomization in the whole process of production, packing, shipment, etc.

L. WHAT IS A SATISFACTORY 'PROBABILITY OF ACCEPTANCE'?

The stringency of a plan is measured by the probability of accepting lots in which a particular proportion of sample units is defective. A relatively lenient 3-class plan suggested in this book ($n = 5, c = 3$) accepts a lot with a 5% proportion of defective units and 30% marginal units on about three occasions in four ($P_a = 0.75$). The most stringent 2-class plan ($n = 60, c = 0$) would accept lots with the same proportion (5%) defective on about one occasion in 20 ($P_a = 0.05$), or lots with 0.5% defectives on about 19 occasions in 20 ($P_a = 0.95$).

It might at first be supposed that the former plan provides no worthwhile protection at all; and that even the latter large-scale examination provides very little. This supposition is negated as soon as one considers the implication of such figures in practice. A probability of acceptance on three occasions in four ($P_a = 0.75$) means that one lot in every four will be decisively rejected, a loss clearly serious enough to compel a manufacturer to reduce the microbial content in his product to a level well below the limit(s) set in the test employed. Even the rejection of one lot in 20 ($P_a = 0.95$) should suffice to have this effect.

Nevertheless, the immediate protection conferred on the consumer as regards a particular lot is seriously limited when using practicable numbers of sample units; hence the recommendations to use large values for n when a direct hazard is recognized (as, for example, in Chapter 6, Section B on *Salmonella*).

The sampling plan, however, has also to be considered in the light of the associated criteria. Consider the example of a moderate health hazard, such as the presence of *Staph. aureus*. If the microbiological limit m were placed at a numerical level well below that likely to represent hazard (*M*), one could frequently accept lots containing a high proportion of marginally acceptable units, which would require only a lenient sampling plan. If, however, *m* were placed nearer the hazardous level, one would wish to accept such lots infrequently, requiring the use of a relatively stringent sampling plan. If *m* were at a hazardous level, one would not wish to accept lots containing any units exceeding that level, and 2-class plans of high stringency would be required. Adjustment is apparently possible by choosing limits known by experience to be associated with safety. When this is done, even though a high proportion of lots with substandard units will be accepted, the probability of consuming food that would cause illness is kept low.

Where more than one test is applied, the stringency of the examination is increased, to a degree depending on whether each test is independent of the others. With completely independent tests, the number of lots rejected would be the sum of the number of lots failing to meet the separate acceptance criteria for each. With completely interdependent tests, no change in acceptance would result. This suggests, for example, that where there is an adequate direct test for a pathogen, there is no additional advantage in testing also for other 'indicator' organisms assumed to indicate the presence of that pathogen. In practice there is far from complete interdependence between different tests, and hence, in general, multiple testing imparts greater stringency.

Where severe hazards are concerned, so that even an occasional *C. botulinum* represents an intolerable risk (e.g., in some canned foods), no conceivably practicable sampling plan could afford sufficient immediate protection to the consumer. In such cases safety must be sought in procedures other than direct bacteriological examination (see Chapter 7).

M. SELECTING *n* AND *c*

The choice of *n* and *c* varies with the desired stringency (probability of rejection) and thus with the cases in the grid of Tables 6 (p. 43) and 10. For stringent cases *n* is high and *c* is low; for lenient cases *n* is low and *c* is high. As *n* decreases with the attributes plans proposed in this book the chance of acceptance of bad lots increases. This fact must be taken into account if the number of sample units (*n*) exceeds the analytical capability of a laboratory and *n* is reduced.

Table 10 shows sampling plans which demonstrate this principle and is intended as a general guide to selecting plans. The choice of n is usually a compromise between what is an ideal probability of assurance of consumer safety and the work load the laboratory can handle. Consider first the nature of the hazard, then decide the appropriate probabilities of acceptance for the hazard in question. Figure 3 (p. 66) has illustrated the decision steps needed to choose between sampling plans.

The stringency of any of the sampling plans can be increased by adjusting c and/or n. In this way pressure can be brought to bear upon the hygienic practices, nature of purchasing specifications, severity of processing, and extent and nature of the quality control practised within the food industry concerned. The desirable effect should be carefully weighed, and the decision made should be known and understood by producer, manufacturer, distributor, and control agency alike.

In summary, the final judgment on which a sampling plan is to be based should involve the relative weight to be placed on the foregoing microbiological, epidemiological, and ecological factors as well as the statistical probabilities of acceptance or rejection desired and economic considerations arising within the laboratory (e.g., the nature of its physical facilities, equipment, and personnel). Some degree of subjective judgment is unavoidable at present, because data adequate to allow fully objective decisions to be made are not available. Under such circumstances, individual judgments can vary widely, and if applied to international plans could give rise to confusion. For this reason the Commission has spent much time in collaboration with international consultants to reach the collective agreements expressed in the plans recommended.

N. INVESTIGATIONAL SAMPLING

Although the chief concern of this book is the use of sampling plans for routine examination, circumstances involving a known or potential microbiological problem require 'investigational' examination. Most of the plans recommended in Part II are designed for routine examinations. Presumably they would fulfil their safeguarding role whether or not the food was from a known manufacturer, on the assumption that the routine examinations would detect severe fault.

Specific investigational plans are not proposed, because the circumstances warranting the use of such plans are so variable. The scale of any increased analytical effort will depend critically upon the com-

TABLE 10

Suggested sampling plans for combinations of degrees of health and conditions of use (i.e., the 15 'cases')

	Conditions in which food is expected to be handled and consumed after sampling, in the usual course of events[a]		
Degree of concern relative to utility and health hazard	Conditions reduce degree of concern	Conditions cause no change in concern	Conditions may increase concern
No direct health hazard Utility, e.g. shelf-life and spoilage	Increase shelf-life Case 1 3-class $n=5$, $c=3$	No change Case 2 3-class $n=5$, $c=2$	Reduce shelf-life Case 3 3-class $n=5$, $c=1$
Health hazard Low, indirect (indicator)	Reduce hazard Case 4 3-class $n=5$, $c=3$	No change Case 5 3-class $n=5$, $c=2$	Increase hazard Case 6 3-class $n=5$, $c=1$
Moderate, direct, limited spread[b]	Case 7 3-class $n=5$, $c=2$	Case 8 3-class $n=5$, $c=1$	Case 9 3-class $n=10$, $c=1$
Moderate, direct, potentially extensive spread[b]	Case 10 2-class $n=5$, $c=0$	Case 11 2-class $n=10$, $c=0$	Case 12 2-class $n=20$, $c=0$
Severe, direct	Case 13 2-class $n=15$, $c=0$	Case 14 2-class $n=30$, $c=0$	Case 15 2-class $n=60$, $c=0$

a More stringent sampling plans would generally be used for sensitive foods destined for susceptible populations.
b See 'Conclusions,' p. 46, for explanation of extensive and limited spread.

petence and other priorities of the individual laboratory, and on the severity of the anticipated hazard. Under conditions of severe risk, for example if the concern were a critical pathogen (e.g., *Salm. typhi*, *C. botulinum*, *V. cholerae*) or if disease of unknown origin were encountered, it would be normal for the testing agency to undertake studies to the limit of its resources or to organize co-operative studies among several agencies or departments.

'Investigational sampling' is a term used in this book to describe sampling which is related to a suspected or identified problem. Typically, such problems may arise from failure of a lot (or a series of lots) to pass routine inspection or some type of ancillary information. Investigational sampling may be done (i) to confirm that a problem exists, (ii) to assist in describing the nature and extent of a problem, and (iii) to provide information on possible sources of a problem.

It is important to note the difference between investigational sampling and the type of sampling described elsewhere in this book. The attributes schemes discussed previously are intended for *routine* application to lots or consignments presented for inspection at various stages prior to sale to the public. This statement also applies to variables plans (Chapter 8). Two- and three-class attributes plans are ideally suited for regulatory, port-of-entry, and other receiver-oriented situations, where little information is available concerning the microbiological history of the lot. Variables plans require stronger distributional assumptions, but may be applicable to situations such as in-plant quality control, where these assumptions can be properly verified. In either case, the emphasis in this type of routine sampling is to do the minimum amount of work to give the desired degree of security.

Investigational sampling is a much more complex topic than routine sampling since each application is in response to a unique problem. The level of expertise needed to design, conduct, and interpret investigational sampling increases accordingly. While it is not possible in this book to examine this topic in a comprehensive fashion, a brief description of investigational sampling follows. More extensive treatment can be found in most quality-control textbooks.

When a potential problem has been identified, further examination (i.e., investigational sampling) into the nature and extent of the problem is frequently required. Such might be the case if a lot rejection under routine inspection is in dispute, or if the cause of the problem is being sought. The fundamental purpose is not to reach an accept/reject decision on a lot; hence, acceptance sampling plans are inappropriate. For example, routine acceptance sampling may have indicated a contamination problem in a plant via an increased rejection rate of lots. A larger random sample may than be taken from one or two of the rejected lots to confirm that a problem exists and to estimate its extent. Emphasis may then shift to the processing plant. If the decision taken is to investigate the plant, samples could be taken from various sites in the plant in an attempt to locate the source of contamination. It may be determined that certain processing equipment is contaminated and this in turn may lead to further investigation involving, among other things, samples obtained from batches of raw ingredients or from employees. In this way the source may be detected and appropriate measures then taken to correct the existing problem and prevent a recurrence.

Note that this hypothetical investigation has used sampling at various stages. Such sampling need not be random. In fact, random sampling may be the least efficient way to approach such a problem. Biased sampling may be preferred as there is sometimes no need to be 'fair.' The value of biased sampling is easy to envisage. For example, a dried product may have become wet because of container leakage during transport. This may be evident from pack or container appearance, or there may be a history of such problems. In this case it may be appropriate to take samples from the areas most likely to have become wet (e.g., the top packs on a pallet). There are many occasions when this type of non-random problem may be suspected, such as when there has been transit pack damage, pallet collapse, or temperature abuse. Other types of sampling, such as stratified, systematic, cluster, or various combinations of these, may also be used. Generally, once the purpose for sampling has been established, then the most efficient type of sampling applicable to this purpose is chosen. Such topics are beyond the scope of this book and are fairly complex statistically, but an excellent discussion is found in Cochran (1977).

O. TIGHTENED INSPECTION

Tightened inspection may be necessary in certain circumstances. This should not be confused with 'investigational sampling.'

In tightened inspection, the sampling plans given in earlier chapters no longer apply. Tightened inspection usually requires increased sample size n, with a corresponding adjustment of other parameters to make the plan more discriminating and perhaps more stringent (i.e., lower consumer's risk). In 2- and 3-class plans the stringency is increased (for fixed n, m, and M) by making c smaller. When c is already

TABLE 11

Choice of plans for routine or tightened inspection of foods, based on circumstances related to the test food

Warranting routine sampling	Warranting tightened inspection*
A. THE FOOD	
Previous test satisfactory	Prior tests frequently unsatisfactory
Indicator tests give no evidence of severe contamination	Routine tests have revealed severe indicator contamination
Not commonly involved in disease	Of a type frequently involved in disease outbreaks
Not specifically suspect in an outbreak	Food from same manufacturer currently involved in a disease outbreak
	Of a type known to provide serotypes newly causing disease
	Circumstances point to possible involvement in an outbreak
Not primarily destined to sensitive populations	Suspect and destined to a sensitive population
	New type of food or new formulation with some rationale for hazard
	Examination results from different laboratories in conflict
B. THE MANUFACTURER	
Records satisfactory	No records
Sanitation control known to be normally adequate	Known or suspected to exercise unsatisfactory plant-control
	Importer has knowledge of temporary hazardous circumstances at the plant
C. COUNTRY OF ORIGIN	
Known to exercise competent control of plant practice	Endemic or epidemic situations hazardous
Not in areas endemic or currently epidemic for pertinent food-borne disease	Carrier rate high
	Sewage pollution usually severe
	Food control systems primitive

* To be appraised in toto. Not all individual factors would, alone, necessarily warrant tightened inspection.

zero, the sample size *n* must be increased to obtain a more stringent plan (assuming *m* and *M* are fixed).

Tightened inspection may be used to place an additional economic burden on suppliers who have a history of producing or handling lots of consistently poor microbiological quality. It may also be used to

TABLE 12

A. Probabilities of acceptance (P_a) for 2-class plans with $n=10$ to $n=300$, $c=0$

Composition of lot		Number of sample units tested from the population (n)					
% acceptable	% defective	10	20	50	100	200	300
99	1	0.90	0.82	0.61	0.37	0.13	0.05
98	2	0.82	0.67	0.39	0.13	0.02	<
97	3	0.74	0.54	0.22	0.05	<	
96	4	0.66	0.44	0.13	0.02		
95	5	0.60	0.36	0.08	0.01		
94	6	0.54	0.29	0.05	<		
93	7	0.48	0.23	0.03			
92	8	0.43	0.19	0.02			
91	9	0.39	0.15	0.01			
90	10	0.35	0.12	0.01			

B. Probabilities of acceptance (P_a) for 2-class plans with $n=300$ to $n=5000$, $c=0$

Composition of lot		Number of sample units tested from the population (n)					
% acceptable	% defective	300	500	1000	2000	3000	5000
99.9	0.1	0.74	0.61	0.37	0.14	0.05	0.01
99.8	0.2	0.55	0.37	0.14	0.02	<	<
99.7	0.3	0.41	0.22	0.05	<		
99.6	0.4	0.30	0.13	0.02			
99.5	0.5	0.22	0.08	0.01			
99.4	0.6	0.16	0.05	<			
99.3	0.7	0.12	0.03				
99.2	0.8	0.09	0.02				
99.1	0.9	0.07	0.01				
99.0	1.0	0.05	0.01				

TABLE 13

The probability of obtaining one or more defectives in a sample of *n* sample units with proportion *p* of the lot defective

Number of sample units (e.g., cans) examined per sample, *n*	Proportion of the lot that is defective (*p*)			
	0.01	0.001	0.0001	0.00001
200	0.87	0.18	0.02	0
1000	1.00	0.63	0.10	0.01
2000	1.00	0.86	0.18	0.02
3000	1.00	0.95	0.26	0.03
4000	1.00	0.98	0.33	0.04
5000	1.00	0.99	0.39	0.05

SOURCE: Codex Alimentarius Commission (1985). ALINORM 85/16, Appendix II. Draft Code of Hygienic Practice for Processed Meat and Poultry Products, Annex C: Procedures for Investigational Microbiological Examination of Meat Products in Hermetically Sealed Containers

replace routine sampling plans when the degree of concern surpasses that given in Table 10. Some such conditions are presented in Table 11, which compares the circumstances warranting use of routine or tightened inspection. For example, food destined for susceptible populations might be subjected to more stringent plans, particularly if there is concern that a lot might contain a relatively small proportion of units which pose a severe health risk. The 2-class plans shown in Table 12 give some idea of how such plans might be effective. This is illustrated in a different way in Table 13, where the probability of detecting one or more positive samples in batches with very low proportions of defectives is given. Thus, if one unit in every 1000 were contaminated, even a sample of $n = 500$ would result in a 61% chance of accepting the lot; that is, there is only a 39% chance of finding one or more contaminated cans with $n = 500$ when the contamination rate is 0.1%. Clearly, such a plan would be unacceptable for highly toxic contaminants.

It is unlikely that the sampling plans of Table 12B will be used for microbiological testing because of effort and cost. On some occasions it is more effective to carry out simple non-microbiological tests on a large number of sample units. For example, the options for testing a batch of canned food for microbial contamination may be full microbiological examination of can contents or the use of a simple test such as pH measurement to identify contaminated or potentially hazardous cans. The first option is very efficient at identifying contamination but will not allow for a large number of examinations. The measurement of product pH is less likely to identify a contaminated product in which growth has taken place, but large numbers of containers can be examined. As contamination is likely to occur at a relatively low rate, it is very likely that the simple test on a large number of cans is the most effective.

REFERENCES

Bergey's Manual of Determinative Bacteriology, 8th ed., edited by R.E. Buchanan and N.E. Gibbons. 1974. Williams and Wilkins, Baltimore

Bryan, F.L. 1982. *Diseases Transmitted by Foods (A Classification and Summary)*, 2nd ed. Centers for Disease Control, Atlanta

Cochran, W.G. 1977. *Sampling Techniques*, 3rd ed. J. Wiley & Sons, Inc., New York, London, Sydney

Halstead, B.W. 1978. *Poisonous and Venomous Marine Animals of the World*, 2nd ed. Darwin Press, Princeton, NJ

ICMSF. 1978. *Microorganisms in Foods. 1. Their Significance and Methods of Enumeration*, 2nd ed. University of Toronto Press, Toronto

NAS-NRC (National Academy of Science / National Research Council). 1973. *Toxicants Occurring Naturally in Foods*, 2nd ed. National Academy of Sciences, Washington, DC

Riemann, H., and Bryan, F.L. 1979 (eds). *Food-Borne Infections and Intoxications*, 2nd ed. Academic Press, New York

Silliker, J.H., Gabis, D.A., and May, A. 1979. ICMSF methods studies. XI. Collaborative/comparative studies on determination of coliforms using the Most Probable Number procedure. *Journal of Food Protection*, *42*, 638–644

Skerman, V.B.D., McGowan, V., and Sneath, P.H.A. 1980. Approved lists of bacterial names. *International Journal of Systematic Bacteriology, 30*, 225–420

Véron, M., and Chatelain, R. 1973. Taxonomic study of the genus *Campylobacter* Sebald and Véron and designation of the neotype strain for the type species, *Campylobacter fetus* (Smith and Taylor) Sebald and Véron. *International Journal of Systematic Bacteriology*, *23*, 122–134

6

Sampling plans for situations involving direct hazard from pathogens

Judgments about hazards from pathogens are based largely on bacteriological analysis, but the results of such analysis, and conclusions derived therefrom, depend on the sensitivity of the method used.

Early methods of detecting *Salmonella* in faeces or blood of a diseased person were not really suitable for food analysis. When more sensitive or discriminating media were used, some foods previously thought to be generally free of *Salmonella* frequently proved to be contaminated. Highly sensitive methods have not yet been developed for *Vibrio cholerae*, *Shigella dysenteriae*, several zoonotic bacterial pathogens, or viruses in foods. Hence, opinions on frequency of contamination by such species may need review as more sensitive procedures become available or if more 'durable' strains arise, (e.g., the El-Tor strain of *V. cholerae*).

The sensitivity of examination for *Salmonella* has been improved further by increasing the quantity of material actually examined. Previously, it was customary to examine samples of food weighing 1 g, or 10 g; now it is usual to examine 25 g, or 50 g, occasionally even more. The importance of this change will become evident below.

A. THE IDEA OF ZERO TOLERANCE

In the past, it has been customary to specify that pathogens 'shall be absent' from foods, and the Commission proposed that for them no numerical tolerance should be expressed. At least three considerations challenge this ideal:

(i) No feasible sampling plan can ensure complete absence of a particular organism (Chapter 5, Section L). Even when $c = 0$ (i.e.,

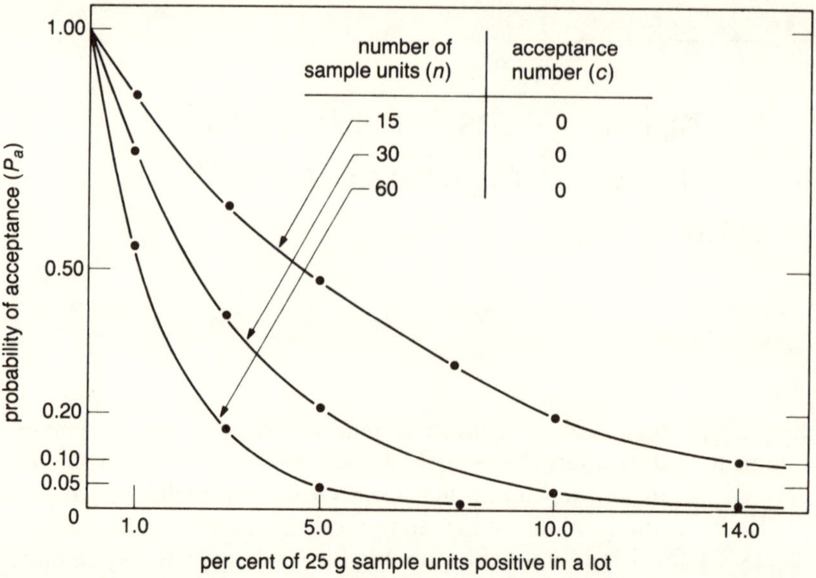

Figure 5 Relations between the sampling plan and the degree of security conferred – OC curves for three sampling plans proposed for *Salmonella*

the lot will be rejected if the pathogen is found in any sample examined), it cannot be guaranteed that the lot is completely free of the organism, no matter how large the number of sample units n. However, what can be found is the probability of acceptance (P_a) for lots of various qualities, as a function of the given n and c. (For example, $n = 60$, $c = 0$ has $P_a = 0.5$, i.e., one chance in two of being accepted, for lots in which 1% of the sample units are defective, i.e., $p = 0.01$; see Fig. 5.)

(ii) Plans in which $c = 0$ are not necessarily the most exacting. For example, if one sets a limit of 5% defectives in a lot, the plan $n = 95$, $c = 1$ will accept 'bad' lots less often, and 'good' lots more often, than will $n = 60$, $c = 0$. In other words, the former plan is the more discriminating, even though it admits the presence of the pathogen. Preference for $c = 0$ is influenced by the wish to emphasize that absence is the desired objective (although it cannot be guaranteed), and by the knowledge that pathogens such as *Salmonella* may be found by direct examination of certain foods and, having been demonstrated, are difficult to overlook.

(iii) It is not yet commercially possible to market some foods completely without pathogens. For example, most raw meats and poultry contain *Salmonella* at relatively high frequency, and no immediate practical solution is in sight. A sampling plan adjusted to this situation would seem more realistic and satisfactory than one based on an unmeasurable ideal of complete absence (i.e., zero tolerance).

The need for compromise

The foregoing illustrates the difficulty of trying to make a rational compromise between a desire to eliminate pathogens completely to protect consumers and what have been considered practicable methods of production. In this field, past decisions have been arbitrary, and some of them illogical. Thus, no restriction is currently enforced on the enormous quantities of poultry and carcass meats which often are regularly contaminated with *Salmonella*, though these foods are clearly shown by epidemiological evidence to be major vehicles for outbreaks of salmonellosis. In contrast, drastic and costly action has been taken on finding *Salmonella* in sugar confectionery and chocolate, though few outbreaks due to such products have been reported. The risk entailed by the regularly contaminated foods is probably as great as that entailed by an operator who is an infective carrier and whose detection would lead to his removal. Therefore, sampling procedures are needed which remove the more highly contaminated lots without completely strangling production.

To encourage a realistic choice of sampling plans for hazardous pathogens, and because of the current international effort to control *Salmonella* in foods, the problem is considered in further statistical detail with particular reference to *Salmonella*. Similar considerations might conceivably be applied, for example, to the *Arizona* group, *E. coli*, or *Shigella*.

B. SAMPLING PLANS FOR *Salmonella*

Any member of the genus *Salmonella* presents some degree of hazard to human health. Some, like *S. gallinarum* and *pullorum*, are of minor significance for humans; others, notably *S. typhi*, are a severe hazard. But the majority, to which this discussion relates, represent a moderate direct hazard which, through cross-contamination and multiplication, can become widespread readily via foods.

The numbers and incidence of these organisms can and should be low. Foods entailing a minimal disease hazard are examined for *Sal-*

monella only when there is unusual cause for concern. Such examination is needed to decide whether a product is acceptable in any of the following circumstances: (a) the food is of a kind commonly involved in salmonellosis; (b) the food is intended for a sensitive population (see p. 51 and Table 8, p. 55); (c) there is no knowledge of the manufacturing processes, or of the bacteriological record of earlier consignments; (d) there is special reason to suspect that the product of a particular factory may contain *Salmonella*.

Because most food control efforts should be directed to the areas of greatest risk, foods are assigned to categories according to the degree of hazard they present (see Tables 6 and 10, pp. 43 and 74). To each category is attached an appropriately stringent method of deciding the acceptability of the finished product. Case 10 recognizes a relatively mild hazard such as exists when a food may contain *Salmonella* but will be used by ordinary consumers after cooking (e.g., red meats, poultry). Case 12 expresses a greater risk, where the conditions of use may favour spread and/or multiplication, and hence needs more stringent sampling. Case 11 is intermediate. Because absence of *Salmonella* is the preferred criterion of acceptance ($c = 0$), the stringency of examination is normally increased by raising the number of sample units n (see p. 72). Table 14 shows recommended sampling plans for the various categories, with degrees of stringency related to hazard expressed by cases.

Our recommendations in Table 14 are similar in principle to those of the report of the *Salmonella* Committee of the NAS-NRC (1969), in that both increase the stringency of the sampling plan with increase in the degree of hazard caused by the conditions of food use. An obvious difference is that the present proposals involve less stringent plans. For this there are two reasons:

(i) The sampling plans proposed by NAS-NRC were not intended for routine examinations, but for use when the question arises whether a product contains *Salmonella*. Routine examinations will from time to time raise this question, and the NAS-NRC plans were intended for the special investigation it entails. For such investigations, the Commission agrees (Chapter 5, Section 0) that plans are desirable in which the degree of stringency is as great as the NAS-NRC recommended (see Table 14).

(ii) If a few highly pathogenic types such as *S. typhi* are regarded separately, the remaining *Salmonella* types which occur commonly in foods (to which Table 14 refers) usually involve mild to moderate disease risks. The NAS-NRC report does not make a similar distinction

TABLE 14

Stringency of sampling for *Salmonella* for foods with different categories of disease hazard (all 2-class plans with $c = 0$)

ICMSF				NAS/NRC[a]	
		n values[b]			
Hazard[c] in use	Case[c]	Normal[c] routine	Special[d]	Category	*n* values[b]
Reduced	10	5	15	B	13
Unchanged	11	10	30	C	29
Increased	12	20	60	D	60

a Based on the Report of the *Salmonella* Committee US NAS/NRC (1969).
b n = number of sample units.
c Compare Tables 6 and 10 (pp. 43 and 74).
d E.g., where the food will be eaten by susceptible consumers (Table 8, p. 55), or for tightened inspection (Table 11, p. 77).

and considers all types of *Salmonella* together, implying a high degree of risk. The Commission agrees that where circumstances (e.g., unusual susceptibility in the probable consumers) create a higher risk, sampling plans of higher stringency are desirable (compare p. 51; and Table 8, p. 55).

Figure 5 gives the operating characteristic curves which express the relation between the degree of security conferred and the sampling plans proposed. For instance, consider the curve for the plan $n = 30$, $c = 0$. If the proportion of sample units defective (p) is 0.05 (i.e., 5% are defective), the probability of acceptance (P_a) is about 0.25, i.e., 1 in 4; hence 3 out of 4 such lots would be rejected. Suppose the units are 25 g samples – a realistic supposition with a food like dried or frozen egg. They are defective if examination shows them to contain any *Salmonella*. Hence, if there are 5% of defective units (i.e., 1 defective in 20) there is, on average, *Salmonella* contamination in 20×25 g = 0.5 kg of food. That is to say, the plan $n = 30$, $c = 0$, with analytical units of 25 g, will reject 3 out of 4 lots in which the incidence of contamination is about two analytical units positive for *Salmonella* per kg.

To reduce laboratory effort while maintaining the stringency of the sampling plan, it would be helpful to combine either groups of analytical units or the enrichment cultures therefrom, using plans in which the presence of a single positive rejects the consignment. A laboratory investigation undertaken for ICMSF (Silliker and Gabis, 1973) has indicated that different sized analytical units drawn from a single well-mixed sample unit of dried food give very similar results. A subsequent investigation (Gabis and Silliker, 1974) using foods of high moisture content (including eggs, poultry meat, meat, and meat products) shows comparable reliability in the use of composite analytical units; for example, examination of three composites of 20 × 25 g analytical units has given the same result as examination of all 60 analytical units individually (the total amount examined being 1500 g in both instances). This finding suggests that considerable reductions may be made in the cost of laboratory analyses by compositing analytical units. Alternatively, such compositing offers the prospect of large increases in the number of analytical units, *n*, with corresponding increase in the stringency of examination, without correspondingly increasing laboratory effort. No reduction in the cost of collecting the necessary number of sample units can be expected, since analytical units to be composited would still have to be chosen randomly.

C. PROBLEMS IN THE IMPLEMENTATION OF STRINGENT SAMPLING PLANS

(a) *Conceivable alternative plans*

As indicated above, for a pathogen like *Salmonella* there are strong reasons for adopting sampling plans in which $c = 0$:

(i) It is objectionable to some people to admit the presence of a recognized pathogen.

(ii) There is the very great technical advantage of combining many analytical units for bacteriological examination in systems with $c = 0$, because a single positive decides the outcome.

(iii) The plan with $c = 1$ (or more) would always require the examination of (many) more analytical units, for equal probability of acceptance, than a plan with $c = 0$.

(iv) Even if the indicated probabilities are the same, their implications are not. For example, when a lot is accepted with the plan $n = 60$, $c = 0$, it is possible (though not certain!) that the lot may perhaps not be contaminated at all. But using the plan $n = 95$, $c = 1$ and accepting one positive, it is known that the lot certainly is contaminated (though the probability that a lot with 5% of analytical units contaminated would be accepted is in both instances $P_a = 0.05$).

(b) *Fallacious procedures*

It might be suggested, for example, that, if the plans $n = 60, c = 0$ and $n = 95, c = 1$ provide equivalent probability (for lots having 5% of units defective), an operator finding one positive in 60 analytical units might then proceed to examine another 35 (total $n = 95$) in the hope of clearing the lot if all the latter proved negative. But such a procedure is in reality a two-stage plan $n_1 = 60, c_1 = 1$, plus $n_2 = 35, c_2 = 0$, which has a greater probability of accepting an unsatisfactory lot than $n = 95, c = 1$. In fact, this probability P_a is 0.07, compared with 0.05 for the one-stage plan. Although in this example the difference is not great, there are situations where such two-stage procedures cause more serious error. Where two-stage sampling plans are actually being used, their OC curves should be computed and the resulting probabilities of acceptance evaluated.

A similar fallacy can arise where a plan requires a large number of analytical units, which would be unusually costly. Suppose the plan is $n = 95, c = 1$, but for economy a group of only 20 units is tested initially. If one unit should fail to comply in this initial group, an analyst might examine the remainder (in this case 75) with the idea that if he does not find any defectives in the second group, he may ignore the first and concentrate on the finding of no defective out of 75 units. Nevertheless, the plan actually corresponding to the tests made is $n_1 = 20, c_1 = 1$ and $n_2 = 75, c_2 = 0$. No justification exists for 'preferring' the results from the second series. Similarly, if the plan were $n = 60, c = 0$, and the analyst felt uncertain about accepting his initial finding of a defective unit in the first 20 analytical units examined, and if the product were accepted on the basis of the second set of results, the actual plan applied would be the double sampling plan $n_1 = 20, c_1 = 1; n_2 = 40, c_2 = 0$, which actually accepts more defective lots than would $n = 60, c = 0$.

D. RELATION TO CURRENT COMMERCIAL PRACTICE

Stringent sampling plans like those just suggested represent an ideal not yet attainable for products like poultry or carcass meats, which are found to be frequently contaminated by *Salmonella* when specially sensitive methods of detection are used. At present it is often recommended to use procedures like rinsing or swabbing of the whole carcass, where in effect the analytical unit is most of the carcass surface and internal cavity, and a positive may correspond to levels

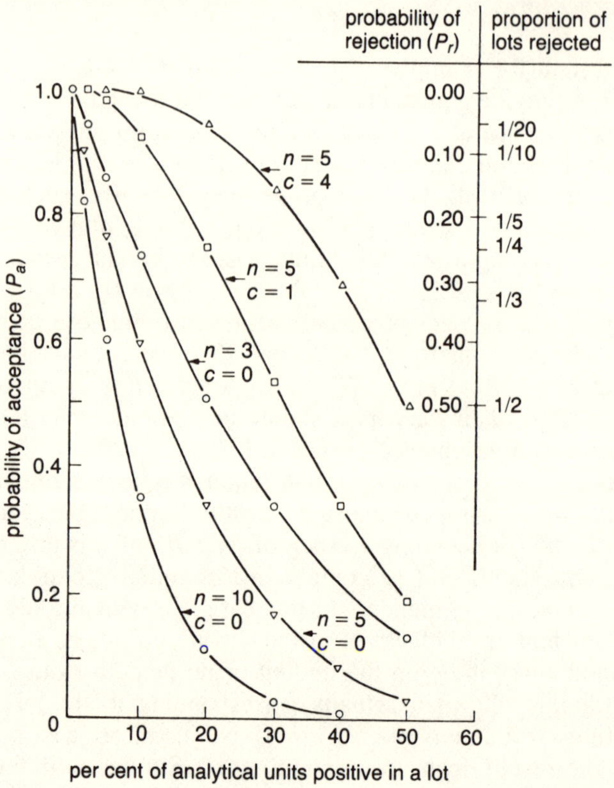

Figure 6 Operating characteristic curves for sampling plans with small numbers of analytical units (data collected from Tables 2 and 3, pp. 34–35)

of contamination of the order of one *Salmonella* cell per kg for poultry, or even one per 100 kg for carcass meats. Using such criteria, one finds slaughter lines producing carcasses found to be contaminated with a frequency of 10% or more. For lots with 10% defectives (i.e., $p = 0.10$), the relations shown in Section A of Table 15 are valid. The sampling plan $n = 60$, $c = 0$ would reject virtually all such material; $n = 5$, $c = 0$ would reject one-third of it (Fig. 6). Even the plan $n = 5$, $c = 1$ would reject 1 in 12 lots, a commercially severe rate of rejection. If, as a result of rejections on the basis of $n = 5$, $c = 1$, the quality of lots improved to the point where only 2% instead of 10% of carcasses were found to be contaminated, the relations would be as in Section B of Table 15. If 2% of carcasses are contaminated,

TABLE 15

Comparison of some operating characteristics of several sampling plans relevant to examination for pathogens (data collected from Tables 2 and 3, pp. 34 and 35)

	A. For lots with $p = 0.10$, i.e., 10% of samples contaminated			B. For lots with $p = 0.02$, i.e., 2% of samples contaminated		
Sampling plan	P_a[a]	P_r[b]	Approximate proportion of lots rejected	P_a[a]	P_r[b]	Approximate proportion of lots rejected
$n = 60, c = 0$	<0.005	0.995	199/200	0.30	0.70	2/3
$n = 10, c = 0$	0.35	0.65	2/3	0.82	0.16	1/6
$n = 5, c = 0$	0.59	0.41	1/3	0.90	0.10	1/10
$n = 3, c = 0$	0.73	0.27	1/4	0.94	0.06	1/16
$n = 5, c = 1$	0.92	0.08	1/12	1.00	<0.005	<1/200

a P_a means probability of acceptance of lot.
b P_r means probability of rejection of lot.

the indicated average concentration of *Salmonella* cells is of the order of one per 50 kg for poultry, and one per 5000 kg or more for beef. If a 2% frequency of contamination were still thought to be too great, a similar pressure for further improvement (rejecting about 1 in 10) could be reintroduced by changing the sampling plan from $n = 5$, $c = 1$ to $n = 5$, $c = 0$.

The above illustrates how, by periodical adjustment of the stringency of the sampling plan, a steady pressure to improve could be applied without placing catastrophic restrictions upon the supply of the commodity at any time.

The sampling plan $n = 5$, $c = 1$ has P_a = about 0.65 for lots with $p = 0.24$ (see Fig. 6); that is this plan would reject 1 out of 3 lots in which one-quarter of the units were contamined. If these units were poultry carcasses weighing about 1 kg, one-quarter contaminated corresponds to a level of contamination of the order of one *Salmonella* cell per 4 kg of food; while with beef carcasses, the figure would be of the order of one per 400 kg. This kind of situation is very different from the examination of analytical units of order 25 g (e.g., of egg). The plan $n = 60$, $c = 0$ rejects about one-third of lots in which 0.7% of analytical units are defective, which (these being 25 g analytical units) corresponds to a minimum level of contamination of about 1 cell per 3 kg. That is to say, the plan $n = 60$, $c = 0$, applied even to 25 g analytical units (e.g., of egg), actually provides less protection than the plan $n = 5$, $c = 1$ applied to whole-carcass samples of poultry of effective weight about 1 kg; even the latter plan would reject a substantial proportion of current production of poultry carcasses.

In such circumstances, the immediate application of the more stringent sampling plans to poultry, and still more to larger carcasses, seems at present unrealistic. The application of even so loose a plan as $n = 5$, $c = 1$ would immediately require a substantial improvement on current levels of contamination; and that improvement could then be continued by proceeding to plans of successively greater stringency. It is important to understand the difference between the sampling plans for some raw meats and those for other foods as regards *Salmonella*, and to note the effect of the size of the analytical unit on stringency of examination, as mentioned at the beginning of this chapter.

REFERENCES

Gabis, D.A., and Silliker, J.H. 1974. ICMSF methods studies. II. Comparison of analytical schemes for detection of *Salmonella* in high-moisture foods. *Canadian Journal of Microbiology*, *20*, 663–669

NAS-NRC (National Academy of Sciences / National Research Council). 1969. *An Evaluation of the Salmonella Problem*. Publication No. 1683, NAS-NRC, Washington, DC

Silliker, J.H., and Gabis, D.A. 1973. ICMSF methods studies. I. Comparison of analytical schemes for detection of *Salmonella* in dried foods. *Canadian Journal of Microbiology*, *19*, 475–479

7

Control at source – the hazard analysis critical control point approach

Any practicable system of bacteriological examination of a food product in commerce can never provide complete certainty that a desired bacteriological state has been attained: it can indicate this only with a particular degree of probability. Moreover, in certain cases (with workable schemes of examination), this degree of probability is either barely adequate, as in trying to detect salmonellae (see Chapter 6, Section B), or even totally inadequate, as in trying to detect *Clostridium botulinum* in canned foods (see Chapter 5, Section L). In such situations it is always wise to use other safeguards besides microbiological examination of the end-product, and where that examination cannot assure reasonable probability of protection, these other procedures become the principal or only safeguard. Thus, for optimal control of microbiological hazards at source, a more complete system is necessary. One such is the Hazard Analysis Critical Control Point (HACCP) system. ICMSF has reviewed this system and assessed its practical use in developed and developing countries. Their considerations are contained in a WHO report (WHO/ICMSF, 1982). The portion of the report that describes traditional control options and the HACCP system follows.

A. CONTROL OPTIONS

As noted in Chapter 1 three principal means have been used by governmental agencies and commercial organizations to control microbiological hazards of foods. These are (a) education and training, (b) inspection of facilities and operations, and (c) microbiological testing. Effective programs utilize combinations of all three approaches.

(a) *Education and training*

Education and training programs for the control of microbial hazards of foods are directed primarily towards developing an understanding of the causes of microbial contamination, including the survival and/or growth of the contaminants. An appreciation of personal hygiene, community sanitation, and food hygiene should be acquired during primary and secondary education. The extent of training required depends on the technical complexity of the food operations and the level of responsibility of the individuals involved. Broad in-depth training may be necessary for some, such as supervisory personnel; for others, training may relate only to some specific aspects of a food operation. Trained personnel should be able to select and apply control measures that are essential for providing safe products of acceptable quality.

(b) *Inspection of facilities and operations*

Inspection of facilities and equipment and observations of hygienic practices of personnel which are often required by regulatory authorities are commonly used to check adherence to good food-handling practices. Such practices may be those considered essential by the inspector, or they may be specified in various advisory or mandatory documents, such as Good Manufacturing Practice (GMP) guidelines, codes of hygienic practice (such as those developed by the Codex Alimentarius Committee on Food Hygiene), or food control laws, ordinances, or regulations. Unfortunately, such documents often contain vague terms, such as 'satisfactory,' 'adequate,' 'acceptable,' 'if necessary,' 'suitable,' relative to some stated requirement, without specifying what is considered to be in compliance with the requirement. This lack of specificity, or some indication of the relative importance of the requirement, leaves the interpretation of compliance solely to the discretion of the inspector. Lack of discrimination between important and relatively unimportant requirements may result in overemphasis upon unnecessary or relatively minor requirements and thus increase costs without significantly reducing hazards. Also, requirements that are critical to the safety of the product may be overlooked or underestimated.

(c) *Microbiological testing or examination*

Samples of ingredients, materials obtained from selected points during the course of processing or handling, and the final product are sometimes examined for microorganisms. Such sampling and testing assist in determining adherence to good manufacturing, handling, and distribution practices. In some instances, foods are examined for specific pathogens or their toxins (e.g., salmonellae or staphylococcal enterotoxins). More often, however, examinations are made to detect either organisms that are indicative of the possible presence of pathogens or spoilage or specific spoilage organisms. Microbiological criteria (i.e., standards, specifications, and guidelines) that state acceptable numerical limits for microorganisms in foods are useful to government and industry. Principles of sampling and the establishment and application of microbiological criteria have been proposed (FAO/WHO, 1979), and sampling plans, test procedures, and decision criteria (limits) for many foods have been suggested by ICMSF (see Part II, Chapters 10–26).

(d) *A modified approach*

While the above control options are widely applied, singly or in combination, there is little epidemiological evidence of their effectiveness, as the incidence of food-borne disease remains high even in the developed countries that apply these measures. Concern over apparent lack of success in this respect, as well as the need to reduce costs associated with ensuring the safety and quality of foods, has led to the development of a more rational approach based on the Hazard Analysis Critical Control Point (HACCP) system. This HACCP concept, first presented at the 1971 National Conference on Food Protection (USDHEW, 1972), was originally developed for use in food-processing establishments (Kaufmann and Schaffner, 1974), but it has been extended more recently to food-service establishments (Bobeng and David, 1977; Bryan, 1981a, b, 1982; Bryan and McKinley, 1979) and to the home (Zottola and Wolf, 1981).

B. HAZARD ANALYSIS CRITICAL CONTROL POINT (HACCP) SYSTEM

The Hazard Analysis Critical Control Point system consists of: (a) an assessment of hazards associated with growing, harvesting, processing/manufacturing, distribution, marketing, preparation, and/or use of a given raw material or food product;[1] (b) determination of critical control points required to control any identified hazard(s);[2] and (c) establishment of procedures to monitor critical control points.[3] The Commission finds that basically the HACCP system provides a more specific and critical approach to the control of microbiological hazards than that achievable by traditional inspection and quality-control procedures, and it intends to develop this system further in a future publication.

(a) *Hazard analysis*

A hazard analysis consists of an evaluation of all procedures concerned with the production, distribution, and use of raw materials and food products: (i) to identify potentially hazardous raw materials and foods that may contain poisonous substances, pathogens, or large numbers of food-spoilage microorganisms, and/or that can support microbial growth; (ii) to find sources and specific points of contamination by observing each step in the food chain; and (iii) to determine the potential for microorganisms to survive or multiply during production, processing, distribution, storage, and preparation for consumption.

1. In food-processing plants

A hazard analysis should be carried out on all existing products and on any new products that a processor intends to manufacture. Changes in raw materials used, product formulation, processing, packaging, distribution, or intended use of the product should indicate the need for reanalysis of hazards, because such changes could adversely affect safety or shelf-life. Microbiological hazards will vary from one product to another, depending on the raw materials, the processing procedures, the manner in which the finished product is marketed, and its ultimate use. They may vary from one food-processing plant to another pro-

1 'Hazards' include contamination of food with unacceptable levels of food-borne disease-causing microorganisms and/or contamination with spoilage organisms to the extent that hazards occur within the expected shelf-life or use of the product.

2 'Critical control point' is a location or a process which, if not correctly controlled, could lead to contamination with food-borne pathogens or spoilage microorganisms or their survival or unacceptable growth.

3 'Monitoring' is the checking or verifying that the processing or handling procedure at the critical control point is properly carried out.

ducing the same product and therefore must be determined by observations and investigations of the particular processing plant.

Hazards caused by microbiological contamination of raw materials must be evaluated. In some cases, microbiological safety and shelf-life depend almost entirely on the selection of microbiologically suitable raw materials. For example, in the manufacture of dry-blended products which are reconstituted without further heating, processing cannot be relied on to eliminate contamination present in raw materials. Also, the stability of low-acid canned foods is dependent on control of the level of thermophilic spore-forming bacteria in the ingredients.

Food products manufactured from raw materials of animal origin should receive special attention because they are the main source of different food-borne diseases in man (salmonellosis, campylobacteriosis, yersiniosis). Other food products that can be assumed to be contaminated with pathogenic microorganisms are of vegetable origin. Such ingredients must be carefully considered in a hazard assessment of processes that utilize them.

Many of the processes used in food manufacture, such as heat treatment, acidulation, fermentation, and salting, will destroy or inhibit the growth of harmful microorganisms. However, other procedures, such as cooling of cooked products, boning of cooked meats, chilling of cans after sterilization in retorts, and slicing of processed meats, may add harmful microorganisms. Hazards associated with these procedures must be evaluated, and the consequences of failure of processing steps designed to destroy or inhibit harmful microorganisms must be understood. For example, the failure of a starter culture to initiate acid production promptly may permit the growth of staphylococci and enterotoxin production during the manufacture of cheese or fermented meats. Failure to allow for the equilibration of pH in an acidified canned food during pasteurization could result in growth of *Clostridium botulinum* spores. Similarly, failure to vent a retort properly prior to heat-processing could result in 'cold spots,' thus leading to underprocessing and failure to destroy *C. botulinum*.

The physicochemical characteristics of a finished product that influence growth, death, or survival of microorganisms should be identified. These include such factors as water activity, pH, the presence of preservatives, the packaging system, and the gaseous environment within it. If interactions between various physical and chemical agents are relied on for safety, such as a_w and pH, pH and preservatives, packaging and gas atmosphere, fermentation and pH reduction, then these factors must be defined in terms of their influence on the microbial flora during processing, distribution, storage, and use by the consumer.

Hazard analysis should include an evaluation of the potential of the food-processing plant environment as a source of contamination to the finished product. For example:

1 To what extent is there opportunity for cross-contamination between contaminated raw materials and finished goods?
2 Is air movement away from finished goods and toward raw materials?
3 Are there steps, such as the manual handling of products that are eaten without further cooking, where employees could contaminate the finished product with pathogenic microorganisms?
4 Is the cooling water of satisfactory microbiological quality?

2. In food-service establishments

Each phase of food-preparation operations – from taking delivery of foods to serving – should be examined step by step for sources or means of contamination, possibilities of the contaminants surviving heating processes, and likelihood of microbial growth. Particular attention should be given to foods and food-preparation procedures that are known from epidemiological studies to be a hazard (Bryan, 1978; ICMSF, 1980). A few examples follow.

Raw meat, poultry, eggs, fish, and rice are frequently contaminated with food-borne pathogens when they reach food-service establishments. For example, poultry frequently harbours salmonellae which may be spread to surfaces of equipment, to the hands of workers, and to other materials. The possibility of cross-contamination to cooked foods must be checked during hazard analyses. For products eaten raw, such as certain fish (common in Japan), oysters, and clams, hazard analyses must concentrate on chilling practices before and after these products arrive at the establishment.

If frozen turkeys are not completely thawed before cooking, salmonellae may survive the cooking process. Also, the interiors of rolled roasts, meat loaves, and various ground-meat products may contain food-borne pathogens. The significance of such contamination must be considered in any hazard analysis of a food-service operation that offers these products. Therefore the thoroughness of cooking these products must be evaluated.

Food handlers constitute a hazard. Cooked ingredients in potato salad, for instance, can be contaminated by persons during peeling, slicing, chopping, or mixing operations in its preparation. Hazard analysis should therefore include observations of food-handling and hand-washing practices of the kitchen staff.

Epidemiological information indicates that the most important factors contributing to the occurrence of food-borne disease outbreaks are related to operations that follow cooking. For instance, rice becomes hazardous after cooking when it is left unrefrigerated for several hours or stored in large masses in large pots overnight. These conditions may permit growth of *Bacillus cereus* and formation of heat-stable toxin. Hazards intensify as the time between preparation and serving of food lengthens. Hazards analyses must assess conditions after cooking, while keeping food hot, cooling, and in cold storage, and reheating practices.

3. In the home
Home-makers can examine their kitchen environments for hazards only if they are aware of what they are. Such awareness can result only from education; that is, at home during childhood, in school, in special courses on home-making, from publications from various sources, and through experience.

(b) *Critical control points*

A critical control point is a location or a process which, if not correctly controlled, could lead to unacceptable contamination, survival, or growth of food-borne pathogens or spoilage microorganisms.

1. Determination at food-processing plants
Incoming raw materials may constitute critical control points, depending on their origin and use. If one or more steps in a process can be depended on to eliminate harmful microorganisms in a particular raw material, that raw material does not constitute a critical control point. For example, the testing, for *Salmonella*, of eggs used in the manufacture of mayonnaise is not usually a critical control point because most countries require mayonnaise to be so formulated that its content of acetic acid and pH will kill these organisms. Also, if the consumers' use of the product destroys contamination, for example, as in the cooking of raw pork sausage, then inspection of the incoming raw pork does not constitute a critical control point. Microbiological examination of such raw materials only provides an indirect means of evaluating general microbial quality. Organoleptic evaluation will often provide more useful information.

If, however, neither processing techniques nor consumer use can be depended on to eliminate harmful microorganisms from raw materials, then these constitute critical control points. Particularly important in this respect are sensitive raw materials, such as dried eggs and milk which may contain salmonellae, and nut meats which may be contaminated with mycotoxins. If raw materials are not controlled at this point, the harmful microorganisms or toxins they may contain are likely to contaminate finished products, such as chocolate confectionery and beverages that are not heated before consumption.

Raw spices may be heavily contaminated with spore-forming organisms which may lead to a significant loss of expected shelf-life of cooked sausages. Thus the examination of such spices for spore-forming microorganisms constitutes a critical control point. However, spore-forming microorganisms in spices used, for instance, in small amounts at mealtime to flavour food are not a critical control point, because they have no relevance to quality or safety.

Processing time-temperature combinations are frequently the most critical control points. For example, if a heat process is depended on to destroy microorganisms, then the required combinations of processing time and temperature must be carefully established and followed. Similarly, the temperatures at which products are held prior to and during cooling and freezing and the length of time they are held are frequently critical control points.

Among other factors that can adversely affect safety and quality are improper sanitation in the plant and packaging materials. If poor sanitation in a particular process step is likely to affect adversely safety of the finished product this would constitute a critical control point. Products may be subjected to environmental contamination from such sources as air, water, insects, rodents, and personnel. Incoming packaging materials do not usually constitute a critical control point, except in the case of containers used for canned foods, where lack of integrity of the finished package may affect the safety or quality of the end-product.

The critical control points in the manufacture of frozen and canned foods have been described (Peterson and Gunnerson, 1974; Ito, 1974).

2. Determination in food-service establishments and homes

Much of the discussion in the previous section is applicable in principle to food-service operations and homes. Places or points in food-service operations or in homes where foods are handled or stored after cooking are particularly important critical control points. These include the handling of cooked foods, keeping hot, cooling, cold storage, and reheating (Bryan, 1978, 1979, 1980).

(c) *Monitoring*

After analysing the hazards presented by a particular product and identifying critical control points, it is necessary to establish monitoring systems to ensure that these points are under control. Such monitoring may involve only visual inspection – for example, the pre-operational inspection of a temperature recorder to determine that the chart has been properly installed. Similarly, since the safety of boned cooked chicken, for example, may be affected by handling by employees, hand-washing procedures should be observed. Although such observations do not involve measurements, they should be recorded on suitable check-lists. More commonly, chemical, physical, or microbiological tests are used for monitoring.

For each critical control point the appropriate monitoring test must be determined, the procedures documented, and the frequency of testing specified. For example, for critical control points involved in canning operations, evaluation and sampling procedures are available (FDA, 1973a; FPI, 1975).

1. At food-processing plants
In no phase of food-processing is monitoring of critical control points so essential to the safety of the finished products as in the manufacture of low-acid canned foods. Indeed, in the United States, such monitoring is subject to federal regulations (FDA, 1973a, b; FPI, 1975). The extent of such monitoring has been reviewed by Ito (1974). Safety of low-acid canned foods is based on the establishment of heat processes capable of destroying microorganisms of public health significance and spoilage types likely to grow at normal ambient temperatures. Physical and chemical tests are performed during production to ensure that all factors necessary to the application of the established safe processes have been adequately controlled. Numerous checks and tests are made at various critical control points, including the adequacy of ingredient blending, determination of consistency, ratio between solids and liquids, weight of product placed in the container, amount of head space, adequacy of the double-seam, time and temperature during sterilization in retorts, quality of cooling water, and post-processing handling of cooled cans. There is no intent, here, to indicate all of the points that are monitored. Rather, the object is to emphasize the importance of

checking each critical control point to ensure that the established procedures have been properly carried out. It is necessary to specify the method for measuring each parameter, to determine satisfactory limits for each test, and to determine the frequency with which the tests and checks will be employed. The results of these tests must be recorded. If a defect is observed, remedial action should be taken and documented.

Monitoring systems for non-sterilized foods are also often complex and detailed and may involve microbiological examinations, such as in the production of non-fat dried milk or dried egg solids. The prime hazard presented by both products is the danger of post-processing contamination with *Salmonella*. Such products should therefore meet microbiological criteria recommended by the NAS-NRC (1969) and the ICMSF, and those often required by regulatory agencies. Assuming monitoring of the pasteurization process indicates proper time and temperature relationships, the most likely source of contamination of the finished product is the environment. It has been repeatedly shown that when potential for environmental contamination exists, a continuing environmental sampling program is more likely to detect a problem than is finished-product analysis. Accordingly, in such operations well-selected points in the environment that constitute critical control points should be constantly monitored. If salmonellae are detected in samples from such points, negative results of tests on finished products should be interpreted with extreme care. An analogous situation would be presented by a dry-blended product composed of multiple ingredients, each of which had been pre-tested and found negative for *Salmonella*. Here, raw materials, the environment, and the finished product are critical control points which must be subject to constant monitoring.

2. In food-service establishments

Many of the measures for monitoring critical control points described in the previous section are applicable to food-service establishments. Visual inspection, however, is the usual approach. Inspection forms developed for the purpose of hazard analysis, which are useful for recording information observed during monitoring, include those related to the inspection of incoming foods, storage conditions, and each step of preparation of potentially hazardous foods. Certain foods – milk and milk products, canned foods, meat, poultry, shellfish – should come from known safe sources that have been subjected to previous monitoring. These and other foods should be checked when received for integrity of can or other packaging, signs of spoilage, and perhaps temperature. Thermometers in cold-storage rooms should be checked.

Particular attention during inspection should be given to temperature of foods, hygienic practices and techniques of handling foods by workers, whether employees are ill or have infections likely to be transmitted by food, and opportunities for cross-contamination from raw foods to cooked foods.

If the cleanliness of equipment is a critical control point, managers should establish a hygiene maintenance schedule that specifies what should be cleaned, how it should be cleaned, when it should be cleaned, and who should clean it. Daily checks should be made to determine whether the schedule is being followed. The capability of cleaning equipment and the effectiveness of the procedures can be evaluated by checking temperatures, length of washing and rinsing cycles, water pressures, and concentrations of detergents or disinfectants. Furthermore, microbiological monitoring can be done on surfaces of equipment to evaluate the efficiency of cleaning. Physical and chemical evaluations of cleaning and disinfection processes, however, are usually of more value.

Measuring the temperature of foods during preparation and storage is probably the most useful monitoring technique for critical control points in food-service operations. Temperatures of certain potentially hazardous foods, such as poultry and pork products, should be measured at the completion of cooking and reheating, or a short time thereafter (during the period of post-heating temperature rise), to determine whether the interior has reached a temperature at which vegetative cells of food-borne pathogens would be killed. Particular attention should be given to the monitoring of temperatures of cooked foods while kept hot or cooling. Several temperature measurements taken at intervals are necessary to evaluate time-temperature conditions (Bryan and McKinley, 1979).

Microbiological sampling and testing of foods at various steps of processing or of finished products provide additional means of monitoring critical control points. Interpretation of these, however, must be based on the ingredients and all of the previous steps of preparation, heating, and storage.

3. In homes

Monitoring in homes includes the following: observing that certain foods in their preparation actually come to a boil, checking pressures (temperatures) and times during heating of low-acid canned foods, inserting a thermometer into meat and poultry during cooking, and checking that shallow containers are used to store cooked foods in refrigerators, that foods are not left at room temperature for several hours, and that in the case of heat-and-serve items the manufacturer's instructions with regard to storage and preparation of the food are followed.

C. APPLICATION OF HACCP SYSTEM

(a) *Approaches to hazard analysis*

When considering hazard analysis both food-poisoning and spoilage microorganisms are of concern. Knowledge that a food is hazardous may derive from one of two sources. (i) Epidemiological information indicating that a product is potentially a health hazard or is microbiologically unstable may derive from effective surveillance programs that collect data on the incidence of food-borne disease and assess significance. Feedback from marketing sources of a particular product may indicate hazards relating to stability or, on occasion, food-poisoning hazards. From the viewpoint of hazard analysis, epidemiological and marketing information is most desirable, as the assessment is based on factual information. (ii) Technical information may indicate that the product poses a health hazard or is subject to spoilage. Here, reaching a decision with respect to hazard is far more difficult than in the first situation. While accurate data concerning product composition and the influence of processing can be obtained at the processing level, it is often difficult to relate this information to the subsequent effects of storage, distribution, and actual use of the product. This lack of information necessitates additional safeguards in analyses, which frequently lead to over-control.

If an existing product or a new product concept is to be subjected to hazard analysis, a food microbiologist with extensive knowledge of the requirements for the type of product under evaluation should be consulted. For example, the following questions should be considered:

1 What are the conditions of intended distribution and use?
Is the product to be distributed under ambient or cold-storage temperatures?
What is the expected shelf-life both during distribution and storage and in the hands of the person who will ultimately use the product?
How will the product be prepared for consumption?

Is it likely to be cooked and then held for a period of time before consumption?
What mishandling of the product is likely to occur in the hands of the consumer or during marketing?

2 What is the product formulation?
What is the pH?
What is the water activity?
Are preservatives used?
What packaging is used, and is this integral to product stability, (e.g., the vacuum-packaging of fresh meats)?

3 What is the intended production process?
Consideration should be given to those steps that lead to the destruction, inhibition, or growth of food-borne disease or spoilage microorganisms.

Based on answers to the above, and other available information, the expert food microbiologist is able to give a preliminary assessment of the potential hazard(s) involved in the manufacture, distribution, and use of the product. However, it is desirable and in many cases necessary to check the assessment by inoculation of the product with appropriate food-borne pathogens and potential spoilage organisms. The inoculated food must be packaged under intended marketing conditions and then be subjected to tests under expected storage, distribution, and consumer-use conditions. Such tests should include evaluation of the effects of mishandling on product safety and stability. The test protocols, including the nature and size of the inoculum as well as other details, should be under the direction of an expert food microbiologist.

(b) *Approaches to critical control point determination*

Sometimes critical control points are obvious from the hazard analysis. Epidemiological data collected during investigations of outbreaks of illness that occurred in similar places can also be used as a guide. At other times, more extensive research on the food or the process, including microbiological investigations, may be necessary to establish appropriate control points. Of particular value is a determination of temperatures and times at which the product is held during processing or preparation for consumption at an establishment.

Microbiological investigations usually form a vital part of the procedure of selecting critical control points, and should include in-depth investigations of raw materials to establish types and numbers of organisms that may be a hazard in a final product, as well as collection of samples of products and/or materials from the surfaces of equipment coming in contact with the food at various stages during the manufacture or preparation for consumption. Statistically valid sampling procedures should be used and repeated on a number of occasions to obtain a realistic picture of the status at different operations, in order to identify stages in processing and the environment where unrestricted microbial growth may occur. Generally, these can be detected by simple aerobic plate count determinations.

(c) *Approaches to establishment of a monitoring system*

The type of monitoring system depends on the nature of the critical control point under consideration.

1. If a *raw material* is a critical control point, a specification should be set for that raw material detailing the microbiological tests, sampling plans, and limits to be employed. Ideally, the supplier will perform the required tests and ensure that the materials comply with the specifications before the product is delivered to the user. However, it may be advisable for the user to check the consignment upon receipt, particularly if it is from a new supplier. Storage conditions of raw material should be monitored to ensure that the satisfactory quality of the material is maintained until it is used.

2. Monitoring of *process* critical control points may involve microbiological tests but may be best achieved by physical and chemical tests, because the results of these are more rapidly available. There are, however, situations where in-process microbiological monitoring is necessary, as, for example, in the production of highly sensitive foods for infants, children, or malnourished persons. It may also be necessary to monitor the effectiveness of sanitation measures by the use of microbiological tests. In situations where a heat-stable toxin is a potential hazard, the product should be examined for numbers of the toxigenic organisms prior to undergoing a heat process to assess the likelihood of the hazard.

3. *Visual observation*, although it may appear to be a mundane activity, is often the key means of monitoring critical control points. Personnel responsible for such monitoring require considerable training and expertise.

4. *End-product* monitoring by microbiological testing is generally very limited. More often, determination of product attributes, such as

pH, water activity, preservative level, and salt content, will give far more information about safety and stability. There are situations where microbiological examination of the finished product is mandatory, such as certain high-risk foods for *Salmonella*. For this purpose, the sampling plans and analytical procedure recommended in Part II of this book should be followed.

Check-lists should be employed for monitoring critical control points. These should show details of the location of the points, the monitoring procedures, the frequency of monitoring, and satisfactory compliance criteria.

REFERENCES

Bobeng, B.J., and David, B.D. 1977. HACCP models for quality control of entrée production in food-service systems. *Journal of Food Protection*. *40*, 632–638

Bryan, F.L. 1978. Factors that contribute to outbreaks of foodborne disease, *Journal of Food Protection*. *41*, 816–827

– 1979. Prevention of foodborne diseases in food-service establishments. *Journal of Environmental Health*, *41*, 198–206

– 1980. Foodborne disease in the United States associated with meat and poultry. *Journal of Food Protection*, *43*, 140–150.

– 1981a. Hazard analysis of food service operations. *Food Technology*, *35* (2), 78–87

– 1981b. Hazard analysis critical control point approach: Epidemiologic rationale and application to food service operations. *Journal of Environmental Health*, *44*, 7–14

– 1982. Microbiological hazards of feeding systems. In *Microbiological Safety of Foods and Feeding Systems*. ABMPS Report No. 125, National Research Council. National Academy Press, Washington, DC

Bryan, F.L., and McKinley, T.W. 1979. Hazard analysis and control of roast beef preparation in food-service establishments. *Journal of Food Protection*, *42*, 4–18

FAO/WHO (Food and Agriculture Organization / World Health Organization). 1979. *Microbiological Criteria for Foods*. Report of a Joint FAO/WHO Working Group on Microbiological Criteria for Foods, Geneva, 20–26 February 1979. (Document WG/Microbiol/79/1). WHO, Geneva

FDA (United States Food and Drug Administration). 1973a. Thermally processed low-acid foods packaged in hermetically sealed containers. Part 128B (recodified as Part 113), *Federal Register*, *39*, No. 16, 24 January, 2398–2410

– 1973b. Emergency permit control. Part 90 (recodified as Part 109), *Federal Register*, *38*, No. 92, 14 May, 12716–12721

FPI (Food Processors Institute). 1975 *Canned Foods – Principles of Thermal Process Control in Container Closure Evaluation*, 2nd ed. National Canners Association, Berkeley, California

ICMSF. 1980. *Microbial Ecology of Foods*. Vol. 2. *Food Commodities*. Academic Press, New York

Ito, K. 1974. Microbiological critical control points in canned foods. *Food Technology*, *28* (9), 46–51, 84

Kaufmann, F.L., and Schaffner, R.M. 1974. Hazard analysis, critical control points and good manufacturing practices regulations (sanitation) in food plant inspections. *Proceedings of the IV International Congress of Food Science and Technology*, 402–7. Institute de Agroquimica y Tecnologia de Alimentos, Valencia, Spain

NAS-NRC (National Academy of Sciences / National Research Council). 1969. *An Evaluation of the Salmonella Problem*. Publication No. 1683, NAS-NRC, Washington, DC

Peterson, A.C., and Gunnerson, R.E. 1974. Microbiological critical control points in frozen foods. *Food Technology*, *28* (9), 37–44

USDHEW (United States Department of Health, Education, and Welfare). 1972. *Proceedings of the 1971 National Conference on Food Protection*. Government Printing Office, Washington, DC

WHO/ICMSF. 1982. *Report of the WHO/ICMSF Meeting on Hazard Analysis: Critical Control Point System in Food Hygiene*. (Document VPH/82.37) WHO, Geneva, 9–10 June 1980

Zottola, E.A., and Wolf, I.D. 1981. Recipe hazard analysis – RHAS – A systematic approach to analysing potential hazards in a recipe for home preparation/preservation. *Journal of Food Protection*, *44*, 560–564

8

The application of variables plans

A. VARIABLES PLANS

Variables plans for lot acceptance sampling are similar to attributes plans in that they accept or reject a lot depending on a sample drawn from the lot. However, unlike attributes plans, variables plans make full use of the microbial counts rather than ascribing the counts to categories, but for correct application of variables plans the microorganisms must be distributed 'log-normally' throughout the lot (i.e., counts transformed to logarithms conform to a normal distribution). The variables plans discussed in this chapter are based on the sample mean and standard deviation of log-transformed counts. Such detailed microbiological information is rarely available at port-of-entry, so variables plans are generally more useful to producers than to importers.

B. RELATIONSHIP TO ATTRIBUTES PLANS

In Chapter 4 attributes plans are described for use at ports, or similar points, of entry. As described, these plans are particularly useful in this context as they do not make assumptions about the frequency distribution. The plans can also be applied to lots of foods under other circumstances, but if the frequency distribution is known, variables plans may also be applicable. Better discrimination may be achieved under these circumstances with a variables plan.

C. IDENTIFICATION OF AN APPROPRIATE FREQUENCY DISTRIBUTION

If, for example, the frequency distribution of the microorganisms within

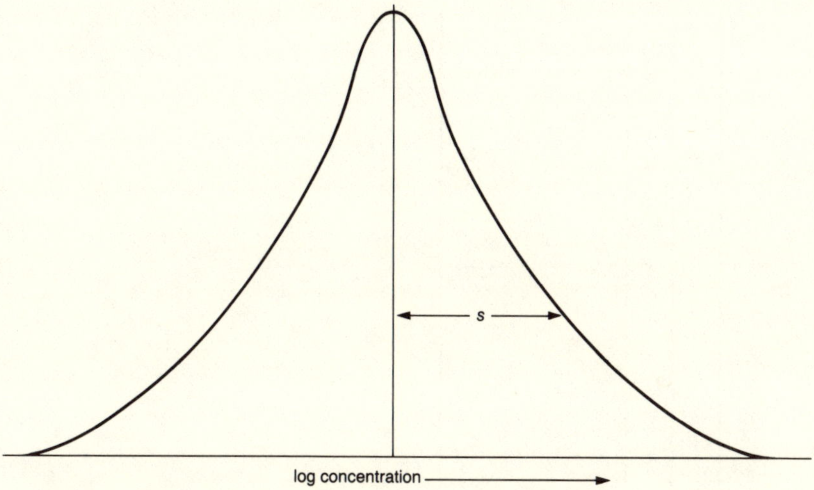

Figure 7 The log-normal distribution

a lot is log-normal, then sampling plans based on the characteristics of this distribution can be used to develop acceptance sampling plans. Microbiological concentration measurements have often been found to follow this distribution in food after log transformation of the data (i.e., if a large number of sample units is analysed, a histogram of log concentration measurements from a lot appears as in Figure 7). When plotting such histograms the interval limits chosen must be equal on a logarithmic scale; that is, 10–31 and 32–99 (1.0–1.49 and 1.50–1.996 on a $\log_{10}$ scale). If the ranges 10–49 and 50–99 were chosen, these would be unequal limits on a $\log_{10}$ scale (1.0–1.69 and 1.70–1.996). If the evidence for a log-normal distribution exists, then it is possible to produce variables sampling plans based on it.

D. OPERATION OF A VARIABLES PLAN

It is necessary to obtain and handle samples and sample units in the same way as for attributes plans. The log-transformation of the concentration measurements is used to compute the sample mean ($\bar{x}$) and standard deviation (s). These two values are then used to make the decision whether to accept or reject the lot. The lot is rejected if

TABLE 16

k_1 values calculated using the non-central t distribution – safety/quality specification (reject if $\bar{x} + k_1 s > V$)

Probability (P) of rejection	Proportion (p_d) exceeding V	Number of sample units							
		3	4	5	6	7	8	9	10
0.95	0.05	7.7	5.1	4.2	3.7	3.4	3.2	3.0	2.9
	0.1	6.2	4.2	3.4	3.0	2.8	2.6	2.4	2.4
	0.3	3.3	2.3	1.9	1.6	1.5	1.4	1.3	1.3
0.90	0.1	4.3	3.2	2.7	2.5	2.3	2.2	2.1	2.1
	0.25	2.6	2.0	1.7	1.5	1.4	1.4	1.3	1.3

TABLE 17

An example of aerobic plate counts for a sample of poultry ($n = 5$)

APC	$\log_{10}$ (APC)	Mean log ($\bar{x}$)	Standard deviation (s)
40,000	4.602		
69,000	4.839		
81,000	4.909	5.039	0.378
200,000	5.301		
350,000	5.544		

$\bar{x} + k_1 s > V$, where V is a log-concentration related to safety/quality limits.

The value k_1 is obtained by reference to appropriate tables and is chosen to define the stringency of the plan for a given number of sample units, n.

E. THE SELECTION OF k_1

Table 16 contains a range of k_1 values for sample unit numbers between 3 and 10. To choose k_1 it is necessary to decide on the maximum proportion p_d of the units in the lot which can have a concentration above the limit value, V. Having selected p_d, the desired probability P, can be chosen, where P is the probability of rejecting a lot which contains at least a proportion p_d above V.

For example, if five sample units are analysed per lot then the k_1 value can be chosen from Table 16. If a lot in which 10% of sample units exceeded V was to be rejected with a probability of 0.95, then the k_1 value 3.4 would be used.

In practice, the two values p_d and P will be selected along with the value V. The scheme then allows n to be selected over the range 3 to 10. The larger n becomes, the lower the chance of rejection of an acceptable lot.

F. SELECTION OF THE LIMIT VALUE V

The limit value, V, is selected by the microbiologist as the safety or quality limit log-concentration. This value is likely to be numerically very similar to the value M used in the 3-class attributes plans (Chapter 4).

TABLE 18

k_2 values calculated using the non-central t distribution – GMP limit (accept if $\bar{x} + k_2 s < v$)

Probability (P) of acceptance	Proportion (p) exceeding v	Number of sample units 3	4	5	6	7	8	9	10
0.90	0.05	0.84	0.92	0.98	1.03	1.07	1.10	1.12	1.15
	0.10	0.53	0.62	0.68	0.72	0.75	0.78	0.81	0.83
	0.20	0.11	0.21	0.27	0.32	0.35	0.38	0.41	0.43
	0.30	0.26*	0.13*	0.05*	0.01	0.04	0.07	0.10	0.12
	0.40	0.65*	0.46*	0.36*	0.30*	0.25*	0.21*	0.17*	0.16*
	0.50	1.09*	0.82*	0.69*	0.60*	0.54*	0.50*	0.47*	0.44*
0.75	0.01	1.87	1.90	1.92	1.94	1.96	1.98	2.00	2.01
	0.05	1.25	1.28	1.31	1.33	1.34	1.36	1.37	1.38
	0.10	0.91	0.94	0.97	0.99	1.01	1.02	1.03	1.04
	0.25	0.31	0.35	0.38	0.41	0.42	0.44	0.45	0.46
	0.50	0.47*	0.38*	0.33*	0.30*	0.27*	0.25*	0.24*	0.22*

* Negative values

G. EXAMPLE OF A VARIABLES PLAN

Table 17 gives the results for the aerobic plate count (APC) analyses of five sample units obtained from a lot of poultry. An appropriate variables sampling plan might be $P = 0.90$, $p_d = 0.25$, with a limit value of $V = 7$. The k_1 value, obtained from Table 16, is 1.7. Applying the formula $\bar{x} + k_1 s$, gives $5.039 + 1.7 \times 0.378$, which equals 5.682. This value is less than the limit value of 7, and the lot is therefore accepted.

H. THE USE OF A VARIABLES PLAN FOR GOOD MANUFACTURING PRACTICE

Food producers often find it advantageous to specify a Good Manufacturing Practice (GMP) standard. It may be possible to apply the variables plan under these circumstances. In order to do this the formula is applied as outlined previously (Section D). The criterion is to accept the lot if $\bar{x} + k_2 s < v$. The k_2 value for the GMP plan is obtained from Table 18. The values P and p_d are selected as before and the appropriate k_2 value is obtained. The limit value, v, will be very similar numerically to the limit value, m, used in the 3-class attribute plan.

For a more extensive treatment of this topic see Kilsby (1982); Kilsby et al. (1979), and Malcolm (1984).

REFERENCES

Kilsby, D. 1982 Sampling schemes and limits. In *Meat Microbiology*, ed. M.H. Brown, 387–421. Applied Science Publishers, London

Kilsby, D., Aspinall, L.J., and Baird-Parker, A.C. 1979. A system for setting numerical microbiological specifications for foods. *Journal of Applied Bacteriology*, *46*, 591–599

Malcolm, S. 1984. A note on the use of the non-central *t*-distribution in setting numerical microbiological specifications for foods. *Journal of Applied Bacteriology*, *57*, 175–177

9

Collecting and handling sample and analytical units

The objective of sampling is to obtain a representative sample of a food and to submit sample units to the laboratory in a condition bacteriologically unchanged from that existing at the time of sampling. All persons concerned with sampling should take appropriate measures to prevent, as far as possible, any contamination of either the food consignment or the sample units and any microbial growth or death within the sample units during transport to the laboratory and during subsequent storage and handling.

The sampling directions given in this chapter are considered good practice and should be followed whenever possible. The directions are those that apply generally, in collecting, labelling, transporting, and storing sample units, and in preparing them for laboratory analysis. It is, of course, difficult to lay down fixed rules to be followed in every situation, and in particular circumstances they may require some modification. Specific sampling directions relating to a given food or class of food are given in Chapters 11–26.

A. SAMPLE UNIT VS ANALYTICAL UNIT

A sample may consist of one or more sample units. A sample unit is the individual portion or container of food randomly taken as part of the overall sample; an analytical unit is that part of the sample unit actually used in an individual analysis in the laboratory. The two may be the same. In many situations, sample units will consist of unopened containers (e.g., hermetically sealed cans, consumer-size packages of frozen foods) holding many times the amount of food needed for the analytical unit.

Each sample unit provides only one result for each test carried out.

The food homogenate prepared from one analytical unit (ICMSF, 1978, pp. 108–111) will often provide enough material for a number of different microbial tests, but occasionally it will be necessary to take two or more analytical units from the same sample unit. For example, when tests for coliforms, staphylococci, and *Salmonella* are being run on the same food, one analytical unit is required for the preparation of the homogenate (to be used for tests for both the coliforms and staphylococci) and one other for *Salmonella* enrichment.

B. GENERAL DIRECTIONS

(a) *Materials*

1 *Sample containers*. Use clean, dry, sterile, leakproof containers such as wide-mouth glass or plastic jars or bottles, stainless metal cans, or disposable plastic bags, whose capacity is adequate for the sample unit desired. Usually they have a capacity of at least 8 oz (227 g). Containers to be reused should be of a quality suitable for repeated sterilizations (see (b) below). Use jars, bottles, and cans with screw-type caps. Insoluble non-absorbent cap liners are necessary; impervious liners made of pressed pulp faced with vinylite are satisfactory. Moulded rubber, plastic, or cork stoppers may be used, but should be covered with an inert material such as aluminum foil or a plastic sheet before they are pressed into the sample container. Disposable plastic bags must be sealed securely after filling so that they cannot leak or become contaminated during normal handling. It is a good practice to place one bag inside another (double thickness) to resist puncture.
2 *Instruments for opening food packages*. Sterile scissors, knives, can-openers, saws, as required, can be used to open food packages.
3 *Sampling devices*. Sterile scoops, spoons, triers, probes, drill bits, tongs, forks, cork-borers, pipettes, and swabs can be used for collecting sample units. Sterile scissors or knives may be needed to cut portions from large items of food.
4 *Labels or markers*. Containers can usually be labelled by marking them with a felt-point indelible marking pen. However, light-coloured waterproof cardboard tags with reinforced eyelet holes and wire or cord ties or gum-backed paper labels or adhesive-backed tape can be used to identify samples. Labels should be big enough to carry all relevant data (see (e) below).
5 *Sterilizing and disinfecting equipment*. In the laboratory, sampling containers or implements should be sterilized in either an autoclave or

an oven capable of reaching a temperature of 170°C. In the field, a pressure-cooker, a steam-chamber, or a hydrocarbon (propane or butane) torch or burner is necessary for sterilization. Disinfecting agents, such as alcohol (95% ethyl or isopropyl), which is subsequently flamed, or hypochlorite solution, which is not a sterilant (made up to 100 ppm and discarded when it falls below 50 ppm), may be used in the field when the above procedures are not practical.

6 *Storage and transport of samples*. A plastic-foam or other insulated container of sufficient capacity is needed for holding and transporting chilled or frozen sample units. Appropriate racks may be needed to secure sample units or prevent leakage or contamination of them during transit. Either ice or plastic or metal containers filled with refrigerant or water and frozen are needed to keep sample units cool. Dry ice is needed to keep frozen sample units frozen.

7 *Cold storage of samples*. In the field, sample units should be stored in either a refrigerator in the establishment from which the samples are being collected or an insulated container with ice or cold containers as described above. Sample units may be chilled first under running water or in ice- or water-baths. In the laboratory, a refrigerator capable of maintaining the sample units at 0–5°C is essential. A freezer or freezer compartment of a refrigerator, depending on the volume of frozen samples received, capable of maintaining samples at −10°C is also required.

8 *Temperature-measuring device*. A device such as a thermometer or thermocouple probe and potentiometer is needed to measure temperatures of food and storage units. Many varieties of digital or recording potentiometers with one or more thermocouple probes are available; these offer more flexibility than thermometers. Metal bayonet (at least 5 inches long), dial-faced (graduated at intervals of not less than 2°C and covering a range of −10°C to 110°C) thermometers have proved useful for measuring food temperatures in the field.

(b) *Sterilizing sampling containers and utensils*

1 Sterilize all sample containers and utensils that are to come in contact with a food. As a rule, sterilize beforehand in the laboratory by one of the following methods:

a Autoclaving at 121°C for 15 minutes. Use this method for all materials containing water and materials likely to be damaged by dry heat (e.g., rubber).

b Exposing glassware, metal containers, and implements and similar materials to hot air (170°C) in an oven for at least 1 hour.

2 Try to bring enough pre-sterilized sampling utensils from the laboratory to collect the desired number of sample units. If additional samples are to be collected after the pre-sterilized utensils have been used, clean (wash, rinse, and dry) used ones at the site and sterilize or disinfect them by using one of the following methods:

a Autoclave them in a small (20–30 litre) pressure-cooker at 121°C for 15 minutes.
b Expose them in a chamber to steam at 100°C for 1 hour.
c Heat them in a small, portable hot-air oven at 170°C for 1 hour (assuming power is available).
d Flame the part of the utensil that will contact food thoroughly with a propane or butane torch or Bunsen burner (taking care not to overheat sampling equipment).
e Immerse them in 95% ethyl or isopropyl alcohol for 3 or more seconds and flame to burn off the alcohol. Repeat this procedure twice more, making sure on each occasion that the flame is out before reimmersing the utensils in the alcohol.
f Immerse them for at least 30 seconds in hypochlorite solution containing not less than 50 ppm of available chlorine or other acceptable halogen sanitizers which are of a concentration equivalent to at least 50 ppm hypochlorite (not a sterilant).

Do not use method *d* or *e* if there is danger of fire or explosion. Methods *b* through *f* may not kill all spores.

(c) *Collecting sample units*

1 Only an authorized person properly trained in the appropriate techniques should collect sample units. This person may be assisted by other persons under his or her supervision.
2 If possible, collect sample units during unloading of a cargo of food (from a ship, truck, airplane) randomly from all parts of the load. Otherwise, collect samples randomly from the 'frame' (see Chapter 3, Section D).
3 If all individual packages are accessible, choose those to be sampled by use of random numbers (see Chapter 3, Section C).
4 When sampling large cartons containing smaller packages, choose the desired number of cartons at random, and from each of these, again

at random, take the desired number of smaller packages. At each stage, use random numbers (see Chapter 3, Section C).

5 Submit sample units to the laboratory in original unopened containers whenever practicable. This will reveal the condition of the product as prepared or altered by subsequent storage and, in the case of samples taken at the retail level, as offered to the public.

6 Take at least twice the amount required for analysis in the laboratory (i.e., in most situations, about 200 grams) to provide a reserve portion in case of accident or later enquiry. Store the reserve under conditions in which its microbial condition will not change appreciably.

7 If the product is in containers too large to be easily transported to the laboratory, transfer representative sample units to separate sterile containers under aseptic conditions. Remove surface contamination on appropriate areas of the food package and the sample container by wiping with 70% alcohol, after removing gross dirt by washing or sponging with 70% alcohol or, in the case of paper containers, by removing the outer ply. Open sealed food packages carefully with a sterile cutting instrument such as scissors or a knife, as appropriate. Use a separately sterilized instrument for each package to avoid cross-contamination. Exercise special care where the risk of cross-contamination is high (e.g., with light powders or pressurized packages). Do not open hermetically sealed cans under field conditions. Submit such cans to the laboratory as sample units.

8 If the food is in bulk, sample consistently from various places within the container(s) rather than, say, just from the top or bottom (unless experience assures complete mixing, or unless sampling from some particular place(s) has become accepted as a result of long experience), or choose the sampling site randomly.

9 When sampling from an outlet or sampling port of a bulk container, allow some material to pass through to 'flush' the port before collecting the sample.

10 Avoid contamination when taking a sample unit from bulk or from a food package and when transferring the material to the sample container.

11 Use an instrument appropriate to the physical state of the food for collecting sample units. For example, triers may be useful for sampling blocks of hard cheese, a drill and bit for frozen products, scoops or hollow probes for dried foods, pipettes or dippers for liquids, and spoons for mixed foods (see recommendations for particular foods in Chapters 11–26).

12 Strive always to obtain a representative sample. Where possible,

agitate liquid, free-flowing, or suspended materials until the product is homogeneous. If feasible, unwieldy brittle material, such as certain frozen liquids, may be placed in a large sterile container (e.g., a strong plastic bag) and broken into smaller pieces with a blunt instrument. If this procedure is not feasible, lay the material on a sterile surface and aseptically break it into smaller fragments using a suitable instrument such as a sterile hammer and sterile chisel. Take sample units from the crushed material.

13 Measure the air temperature of the storage room or shipping vehicle as well as the temperature of the food from which the sample is drawn. Do not insert a thermometer into the food until after the sample has been taken. Disinfect the thermometer or thermocouple between measurements to avoid cross-contamination. In instances where small packages are sent unopened to the laboratory, record the temperature of the food in an adjacent package in the same case or carton. The temperature of boneless moist food or dried foods collected in plastic bags can be taken by folding the filled bag around the stem of the thermometer or sensor of a thermocouple and waiting until the temperature reading stabilizes.

(d) *Number of sample units*

Draw the number of sample units specified in the sampling plan used. See Chapters 11–26 for recommendations for individual foods and cases; also Chapter 3, Sections B and C, for what constitutes a representative sample and recommendations on random sampling, and Section B of this chapter for recommendations on the amount of food to take for each sample unit.

(e) *Labelling sample containers*

1 Label all sample unit containers immediately before or after the sample unit is taken. Fix the label to prevent accidental removal during subsequent handling.

2 Number the sample unit container or write the number on the label. Make a related record including all the information necessary to identify the sample (see (j) below).

3 When the sample unit is taken from a larger container such as a bag or carton, identify this container with the corresponding number of the sample unit, in the event a subsequent examination is required.

(f) *Sealing of sample containers*

In certain situations, such as where the food is the subject of a dispute or is suspect in an outbreak of food-borne illness, it may be necessary to place an 'official' seal on the container from which the sample unit was taken, preferably in the presence of persons authorized to represent all parties concerned. The seal may be glued paper or other material which makes it impossible to tamper with the contents or the label without irreparably damaging it. Identify the seal with the date, sample number, and identifying mark of the person collecting the sample. When, for official purposes, multiple samples are required, treat each sample unit identically.

(g) *Transporting and storing samples*

1 Examine sample units as soon as practicable after they are taken. Otherwise transport and/or store sample units under conditions which minimize the possibility of change.
2 Transport sample units to the laboratory for analysis as rapidly as practicable. If the product is canned (shelf-stable) or is in a dry condition, cooling the sample units is unnecessary but avoid temperatures above 40°C. Canned products considered likely to 'blow' should be refrigerated.
3 If the product is perishable and unfrozen, cool sample units rapidly to 0 to 5°C (e.g., in an ice or a water bath) and maintain this temperature during transport (see (a) above). Do not freeze the sample. In transporting sample units ensure that the containers will be held vertically and that liquid from thawed ice in direct contact with the containers does not contaminate the sample units.
4 Until analysis maintain sample units of frozen products in the frozen state.

(h) *Collecting samples during investigations of outbreaks of food-borne disease*

Sampling during the investigation of disease outbreaks (investigational sampling) may be quite different from routine sampling. Using professional judgment, samples of food or equipment surfaces should be collected at the point of operation considered most likely to yield food-borne pathogens; that is, where contamination, survival, or growth might occur (Bryan, 1980; Bryan et al., 1976).

Check storage facilities for food left from the suspect meal or lot. If none are available, collect sample units of food prepared in a manner similar to that used in the preparation of the suspect food as well as of the corresponding raw food (particularly that of animal origin) and ingredients used.

Collect samples of each component of a food (e.g., meringue and filling for custard pies). These have different ingredients and they could have been prepared by different persons or prepared or stored in different ways prior to final assembly.

If the mode of spread or source of aetiological agent is sought, and if definitive typing of isolates is contemplated, swabs of food contact surfaces and other environmental samples can be taken. When collecting swab samples, rub sterile swabs or sponges randomly over large areas of foods of animal origin and over contact surfaces of utensils and equipment used to process, store, or prepare suspect foods. Sample units should be labelled, sealed, transported, and stored as stated in Sections (e) to (g) above.

(i) *Legal requirements*

Before examining foods likely to be the subject of litigation, it is essential that the legal requirements for sampling be determined and followed.

(j) *Sampling report*

1 Complete a detailed sampling report, signed by the person responsible for collecting the samples, and countersigned by representatives of the other parties concerned, if they are present. This report should contain the following information relevant to each lot in the consignment:

a name and address of the person collecting the samples
b names and addresses of appropriate representatives of the parties concerned
c date, place, and time of sampling
d reason(s) for sampling
e nature of the food
f names of the manufacturer, importer, seller, buyer, as appropriate
g number and size of the units constituting the lot
h number and marking of the lot

i type and identification of the transport vehicle
j origin of shipment
k place of destination
l dates of shipment and arrival of the lot
m number of the bill of lading or contract
n method of sampling (random throughout lot, random throughout accessible units, or otherwise)
o size, number, and reference numbers of sample units
p temperature of the product at the moment of sampling
q means of transporting samples to the laboratory and by whom
r name and address of laboratory analysing samples
s tests to be made (e.g., APC, *Salmonella*, *Clostridium perfringens*)

2 The report should also include information on factors, conditions, or circumstances which may have influenced sampling or be pertinent to the investigation; for example, the condition of the individual packages, breakage, sanitation, humidity.

(k) *Preparing sample units for analysis*

Withdraw a representative analytical unit from the sample unit.
1 Begin the analysis as soon as possible after the sample material arrives at the laboratory.
2 If the sample is frozen, use one of the following methods:
(a) If the frozen material can be easily comminuted, do so and proceed without thawing. (b) Partially thaw sample units in their original containers (or in the containers in which they were received at the laboratory) in a refrigerator at 2 to 5°C.
3 Agitate thoroughly any food that is liquid, suspended, or free-flowing, before opening the sample container.
4 Avoid contamination when opening the sample container. Destroy surface contamination on and near the lid area of the container (or where the opening is to be made, in the case of foods in closed packages) by (a) flaming or (b) wiping with 70% alcohol and air-drying or burning away the excess alcohol, as appropriate.
5 Take special care in opening sample containers or packages of powdered materials to avoid contaminating the working area and other samples. If practicable open such containers in a room separate from the testing area or in a cabinet where the powder can be contained. Avoid sudden movements and keep doors and windows closed.
6 When a product is composed of several components or layers (e.g.,

turkey and dressing, cream pie with meringue topping), it is desirable to determine the microbial population or contamination of each component. Such testing can reveal possible sources of contamination or identify an area where improper processing, handling, and/or storage has occurred. Therefore, sample the parts separately while preventing cross-contamination as much as possible.

7 Observe any abnormal appearance or aroma of the sample units and note observations on the record form.

8 Use aseptic technique to withdraw and transfer samples to a sterile blender or stomacher bag.

9 For preparation and dilution of the food homogenate, follow directions given on pages 108–111 of ICMSF, 1978.

10 Store unused portions of the sample units in a refrigerator or a freezer as in (g) above, for replicate analyses if required.

11 Clean and disinfect all surfaces used.

REFERENCES

Bryan, F.L. 1980. Procedures to use during outbreaks of food-borne disease. In *Manual of Clinical Microbiology*, 3rd ed. ed. E.H. Lennette, A. Barlows, W.J. Hausler, Jr, and J.P. Truant, 40–51. American Society for Microbiology, Washington, DC

Bryan, F.L., Anderson, H.W., Anderson, R.K., Baker, K.J., Matsuura, H., McKinley, T.W., Swanson, R.C., and Todd, E.C.D. 1976. *Procedures to Investigate Food-Borne Illness*, 3rd ed. International Association of Milk, Food and Environmental Sanitarians, Ames, Iowa

ICMSF 1978. *Microorganisms in Foods. 1. Their Significance and Methods of Enumeration*, 2nd ed. University of Toronto Press, Toronto

PART II

SPECIFIC PROPOSALS FOR SAMPLING AND SAMPLING PLANS

10

Introduction: The application and use of criteria

This chapter provides background information on the considerations which have led the Commission to propose microbiological criteria for some commodities and not others. It also indicates how the criteria should be interpreted and applied.

A. FORMAT OF COMMODITY CHAPTERS

In the first edition of this book foods were grouped on the basis of commodities (e.g., milk and milk products) or processes (e.g., frozen foods, dried foods). In this edition the commodities grouping used in *Microbial Ecology of Foods*, vol. 2 (ICMSF 1980), has been followed with two exceptions. These are formulated foods, comprising raw materials from several commodity groups, and low-acid canned products. Criteria for formulated foods will depend on conditions of manufacture, the types of raw materials used, the process, the intended distribution system, and shelf-life. Hence criteria are proposed only where a need has been demonstrated and such information is available (see Chapter 24, Formulated Foods). Sampling plans and microbiological tests are not relevant to the safety of shelf-stable canned foods and are therefore not proposed. Extensive cross-reference is made to the volume referred to above (ICMSF 1980) so that information pertaining to the need for criteria and the appropriate cases may readily be found.

Each chapter discusses the microbiological hazards associated with a commodity group and, based on a consideration of their relevance, may propose criteria. General sampling procedures are dealt with in Chapter 9, but if a commodity requires special sampling procedures these are detailed within the chapter. For methods, reference is made to ICMSF, 1978, or to other appropriate sources.

B. SELECTION AND APPLICATION OF CRITERIA

Ideally the control of microorganisms in foods is at the point of production, processing, or preparation for consumption (see Chapter 7, Control at Source). However, for much food in international trade there is no knowledge of control at source or of the conditions used during processing and handling. Therefore, there remains a need for criteria to assess the acceptability of foods at port of entry.

Before recommending a criterion for a product, the Commission decided that each product must meet the following conditions:

1 The product must be in international trade.
2 There must be good epidemiological evidence that the product, or product group, has been implicated in food-borne disease and/or may have an inadequate shelf-life if Good Commerical Practice (GCP) has not been followed.
3 There must be good evidence, or good reason to believe, that the application of a criterion will reduce the health risk in food and/or effectively assess adherence to GCP.

It is important to consider some of the difficulties in expecting a microbiological examination to portray the true microbiological condition of food. For example, a relatively few sample units may not accurately reflect the true microbiological status of a consignment, particularly if the consignment is composed of several batches of food. In this case, if only one of the batches is defective, sampling a portion of the consignment may not reveal the unacceptable part. The resulting decision, if applied to the whole consignment, would be quite unsatisfactory. The sampling plans proposed in this book should whenever possible be applied to identifiable lots (see definition of a lot, Chapter 3, Section A).

If at port of entry the intention is to safeguard against staphylococcal intoxication, a criterion for *Staphylococcus aureus* would not detect the hazard if the viable cells have died. If epidemiological evidence shows that this product constitutes a hazard, a more direct criterion would then be needed, such as one for thermonuclease or enterotoxin. If a criterion is applied to a finished product it may be inadequate if a destructive process has been applied (e.g., destruction of *Staph. aureus*, but with enterotoxin persisting).

The microbiological status of some foods changes continually during storage and distribution and is taken into account when selecting the 'case.' Thus, control at the source is the ideal, and criteria applied at the port of entry will never be as effective, though there are many instances where their application is effective (e.g., *Salmonella* in dried milk).

C. CHOICE OF ORGANISMS

The ICMSF has attempted to include in criteria those microbes most important in respect of hazard and/or non-compliance with GCP. This choice was based on epidemiological evidence and expert opinion, recognizing limitations of current methodologies.

D. SELECTION OF LIMITS

For 3-class attributes plans it was necessary to establish *m* values (associated with GCP) as well as *M* values (related to the safety/'quality' limit). The latter are based on expert opinion as to the acceptable limit, but the former should be based on firm data obtained from producers and retailers operating according to Good Manufacturing Practice (GMP) and GCP.

Although ICMSF has collected a large amount of data, there is not sufficient information for certain commodities to establish *m* values on this basis. When information was not available, expert opinion was relied upon to establish *m* values for 3-class plans. Where such values could not be derived, 2-class plans are adopted. Appropriate values of *m* for 3-class plans must await the collection of further microbiological data.

The ICMSF has now implemented a pilot computer-based collection and retrieval system for microbiological data and is anxious to obtain data to expand this data base. Readers having relevant data are invited to contact the secretary of the ICMSF, who will be pleased to provide printed forms for confidentially recording data in a format suitable for the system.

REFERENCES

ICMSF. 1978. *Microorganisms in Foods. 1. Their Significance and Methods of Enumeration*, 2nd ed. University of Toronto Press, Toronto

– 1980. *Microbial Ecology of Foods*. Vol. 2. *Food Commodities*. Academic Press, New York

11

Sampling plans for raw meats

World trade in raw meats, both chilled and frozen, continues to be significant in terms of volume despite their potential hazard to health. Total world red meat production in 1979 was more than 78 million tonnes, with about 9% being traded internationally.

A. REASONS FOR SAMPLING

Raw meats are important sources of *Salmonella* and *Clostridium perfringens*, which are often incriminated in outbreaks of food-borne disease (ICMSF, 1980, p. 354). They are also sources of *Staphylococcus aureus*, *Campylobacter fetus* subsp. *jejuni*, and *Yersinia enterocolitica*. Salmonellae are found more often in pork and in veal from young calves than in other meats (ICMSF, 1980, p. 347). In the first edition of this book criteria for salmonellae were included as guidelines to assist control of this organism in raw meat. It was hoped that such guidelines would lead to a reduction in the contamination rate and a drop in the incidence of food-borne disease. However, outbreaks of salmonellosis due to meat have continued with little evidence of diminution. In some instances salmonellosis has been caused by the consumption of raw or inadequately cooked meat products but a more common hazard arises through cross-contamination from raw to cooked meat or other foods, and subsequent time-temperature abuse.

Clostridium perfringens is ubiquitous, and although it occurs on carcass meat, usually in low numbers, it cannot be controlled by any known means. The majority of outbreaks of *C. perfringens* gastroenteritis attributable to meat result from inadequate storage of the cooked product. Prevention involves attention to the time/temperature conditions of cooking and, more important, to hot holding, cooling, and reheating before consumption.

Staphylococcus aureus may occur on raw meat although usually only in low numbers. Contamination by animal strains of *Staph. aureus*, which are thought to have a low enterotoxin-forming potential, is probably of less consequence than contamination from human sources. *Staphylococcus aureus* competes poorly with the normal microbial flora of raw meat and constitutes a health hazard only when this competing flora is restricted and there is temperature abuse of the product.

Campylobacter fetus subsp. *jejuni* is often present in the intestinal flora of healthy animals used for food production (Blaser, 1982). However, the numbers present on red meat are generally low and the organism has only a limited potential for growth or survival on refrigerated or cooked meat. Food-borne outbreaks of *Campylobacter* enteritis appear to arise almost exclusively from eating raw or undercooked food (Blaser, 1982).

The use of microbiological criteria as a control measure for reducing the incidence of disease arising from the consumption of meat carrying *Salmonella*, *C. perfringens*, and *Staph. aureus* was considered by a FAO/WHO Working Group (FAO/WHO, 1979) in light of the Codex 'General Principles for the Establishment and Application of Microbiological Criteria for Foods' (Codex, 1981). According to these principles a microbiological criterion should be established and applied only where there is a definite need and where it is both practical and likely to be effective. Such criteria should form part of, or be related to, a Codex Code of Practice for the particular product. The presence of *Salmonella* and *C. perfringens* on raw meats is generally more a reflection of their incidence in the live animal than of a breakdown of hygiene. In present circumstances the occurrence of salmonellae and the other pathogens in raw meat cannot be entirely prevented by the application of codes of hygienic practice (ICMSF, 1980, p. 347). Also, the extreme variability of distribution of pathogens such as salmonellae in meats prevents the establishment of practical sampling plans which would ensure with any degree of confidence the absence of salmonellae in meats. Consequently the Working Group concluded that the application of limits for pathogenic microorganisms at port of entry was not appropriate in terms of the Codex 'General Principles.'

In view of the importance of *Salmonella* as a cause of food-borne disease and meat as a vehicle for its transmission there is much to be gained from regular in-plant monitoring to establish the extent of the problem in particular areas and to detect increases in prevalence. Products that have a history of being implicated as causes of salmonellosis should be sampled at appropriate points during production and distribution to determine the prevalence of contamination and to trace its source on the farm as well as to look at the effects of travel, holding, and processing of stock. The aim is to collect information that will help to motivate the producer to eliminate salmonellae in the live animal and thus control the problem at its source.

The ultimate solution to the *Salmonella* problem lies in producing *Salmonella*-free animals. The provision of *Salmonella*-free foodstuffs for animals would be a major advance towards this aim (see Chapter 14, Feeds of Animal Origin and Pet Foods). Until this is achieved treatments that destroy salmonellae in the product (e.g., irradiation) may provide some protection for the consumer. In the absence of effective control measures, environmental and line sampling are more useful than testing the end-products. Although improvement of slaughter hygiene is incapable of eliminating contamination with salmonellae, neglect of hygiene can make the situation much worse. A comprehensive monitoring program in the processing plant can provide a check on the prevalence of infection in livestock and the effectiveness of measures of hygiene in controlling spread of contamination.

The FAO/WHO Working Group (FAO/WHO, 1979) concluded also that the number of indicator organisms in meat neither reflects adherence to a code of hygienic practice nor indicates presence or absence of pathogens. Hence criteria based on indicator organisms were not justified for raw meat. These conclusions, and those concerned with pathogens, have been adopted by the Codex Committee on Food Hygiene (Codex, 1981).

B. SAMPLING PLANS

In the virtual absence of studies relating health risks to surface counts of aerobic organisms on carcass meats it is not appropriate to reject consignments of product for public health reasons solely on the basis of high aerobic plate counts. However, aerobic plate counts (APC) can be used to monitor Good Commercial Practice (GCP), and criteria based on such examinations are a valuable aid to in-plant quality control. Microbiological quality control of meat-processing involves development and use of processing methods designed to keep microbial numbers low by reducing contamination and preventing growth. Microbiological monitoring of product and plant at critical control points

TABLE 19

Sampling plans and recommended microbiological limits for raw meat[a]

Product[b]	Test	Method reference[c]	Case	Plan class	n	c	Limit per cm² or gram m	M
Carcass meat, before chilling	APC	118	1	3	5	3	10^5	10^6
Carcass meat, chilled	APC	118	1	3	5	3	10^6	10^7
Edible offal, chilled	APC	118	1	3	5	3	10^6	10^7
Carcass meat, frozen	APC	118	1	3	5	3	5×10^5	10^7
Boneless meat, frozen (beef, veal, pork, mutton)	APC	118	1	3	5	3	5×10^5	10^7
Comminuted meat, frozen	APC	118	1	3	5	3	10^6	10^7
Edible offal, frozen	APC	118	1	3	5	3	5×10^5	10^7

a **Not for use at port of entry** but for in-plant quality control. Refer to Section B for appropriate applications.
b Unfrozen carcasses and primal cuts, swab counts per cm^2; other meats and meat products, per gram.
c Refers to page number in ICMSF, 1978.

can be used to assess the effectiveness of sanitation factors in limiting microbial growth. These counts are compared with APC criteria established for particular products prepared under specific conditions. Sampling plans and microbiological criteria for certain raw meat commodities are listed in Table 19.

The proposals for raw meats are based on limited data collected mainly from production and under a restricted range of circumstances. The criteria have been established as guidelines and some tolerance must be allowed in their application; for example, meat marginally acceptable, or even rejected, by these guidelines could be used in heated products, but would have very limited shelf-life as fresh meat, even under good refrigeration. Meat prepared according to the *Recommended International Code of Hygienic Practice for Fresh Meat* (CAC/RCP 11-1976, FAO, Rome) may on occasion exceed these limits, which should not be used as legal standards. No attempt has been made to distinguish between meat from different species although it has been noted that sheep carcasses normally have a slightly higher level of initial contamination than those of beef and numbers of bacteria tend to be higher again on pig carcasses (ICMSF, 1980, p. 342).

Although there are considerable quantities of horse meat in international trade, criteria have not been included because of lack of data.

The FAO/WHO Working Group (FAO/WHO, 1979) considered that the only point at which APC could be used for chilled meat, to evaluate the hygienic conditions under which it was produced, was at the slaughterhouse, and because of its perishable nature it would not be possible to set APC limits for use in guidelines or specifications for chilled meats at port of entry. This consideration applies to meat that is vacuum-packaged in an oxygen-impermeable plastic film as well as to chilled meat packaged in other ways. Vacuum-packaging and modified-atmosphere packaging bring about changes in the bacterial flora, and the storage life depends more on the nature of the flora that develops than on the numbers of bacteria present after processing.

Case 1 with a 3-class plan for the APC is appropriate for utility tests for general contamination of a product that is normally cooked before consumption. The values for *m* reflect current commercial attainment based on the examination of meat from different sources, mainly at point of production, and are supported by experience that meat with plate counts lower than the value set for *m* usually has not been subject to excessive contamination or undue faulty handling. A normal shelf-life would be anticipated. Overall data from a large number of samples have been analysed but not all products have been tested to a similar degree.

The *M* values for APC are based on experience that meat with APC values in excess of 10^7 is either grossly contaminated or has been exposed to conditions permitting microbial growth to a level not far short of the point at which incipient spoilage can be detected. Immediately after slaughter an *M* value of 10^6 is justified on the basis that counts below this level can be readily achieved under GCP. At this level some tolerance is allowed for further processing which could involve slight increase in microbial numbers.

Microbiological criteria applied to frozen meat should be similar to those applied to chilled meat. The criteria proposed for frozen carcass meat and frozen cuts assume that the meat is sampled in the frozen state and thawed quickly before examination of samples. Spoilage occurs at about the same level; hence the *M* values are the same. Reflecting the small decrease in numbers due to freezing and storage *m* values are 5×10^5 instead of 10^6. Commercially thawed meat usually has a higher count than the frozen product, and the criteria for chilled carcass meat could apply.

The microbiological quality of comminuted meat reflects the situation in the meat from which it is prepared. Counts in general are 10– to 100–fold greater in commercial minced meat than on the equivalent carcass (ICMSF, 1980, p. 370). If scrap meat and trimmings from the outer surface of carcasses are used, the counts in the mince will be higher than if whole cuts are minced. Mechanically deboned meat derived from good-quality raw material can be microbiologically equivalent to or better than minced, manually boned meat provided strict temperature control is exercised, but the process needs careful control in respect to the material being boned and sanitation of the equipment. There have been many proposals for microbiological criteria for minced meat but there is no evidence that a criterion has any relevance to health hazard. Accordingly criteria similar to those for carcass meat are proposed.

There is little information available on the microbiology of edible offals but the considerations are similar to those for carcass meat and the criteria proposed are the same.

C. SAMPLING PROCEDURES

(a) *Carcasses and primal cuts*

For chilled or frozen carcass meat and large bone-in or boneless cuts, as $n = 5$ for all criteria suggested, take subsamples from 5 individual carcasses or cartons of cuts (sample units). If the lot is distributed in several shiploads or freightcars take the sample units from more than one transportation unit.

Because contamination of meat is often very uneven, a number of subsamples should be taken from different parts of the carcass or cut, including areas known to be subject to contamination or particularly favourable for microbial growth. The subsamples from a carcass or carton may be pooled and thoroughly mixed to form a composite analytical unit, or each may be treated as a separate analytical unit.

Bacterial contamination of carcasses or cuts is generally on the surface, and this should be recognized in the collection of subsamples. Because defacing of carcasses or cuts is often economically undesirable, the surface swab technique is recommended for the APC on unfrozen meat. Samples should be taken at sites likely to be contaminated (see, e.g., Roberts et al., 1980). Lamb and pork carcasses should be swabbed at at least two points (leg and brisket), and beef and horse at three (leg, flank, and neck). Sampling of cuts should include the original exterior surface and a cut surface.

For a comparison of swabbing and excision methods see ICMSF, 1980, p. 359.

(b) *Boneless bulk cuts, comminuted meat, and edible offals*

Except for the offals it is usually not possible to identify the original surface. From each of 5 cartons or packages remove a sample unit of about 200 g of tissue, which should comprise subsamples from a number of different parts of the pack. This composite sample unit should be mixed thoroughly and the appropriate analytical unit withdrawn.

(c) *Retail packages*

These include chilled or frozen consumer-portion packed cuts and offals as well as comminuted meat. The sampling procedure entails the examination of 5 packages from the lot.

D. SAMPLE COLLECTION

Remove any wrappings from the carcass or package carefully without handling the meat. Use sterile swabs and templates for taking surface swab samples. Mark the sampling area with sterile metal guides (e.g., 5, 50, or 100 cm^2). Use two sterile swabs to sample each area. Moisten the first swab with sterile peptone water and rub firmly across the exposed area several times in all directions. Use the second swab dry and rub over the same area. Introduce both swabs into a bottle containing 3 or 4 glass beads and an appropriate known volume of diluent (e.g., 0.1% peptone, 0.9% NaCl). Shake vigorously.

To avoid cross-contamination when removing portions of meat, use sterile instruments (knives for unfrozen meat; saws, cleavers, and special drills for frozen meat), and transfer the piece aseptically to sample containers.

Samples of frozen meat should remain frozen during transport to the laboratory. Samples from chilled carcasses should not be frozen.

E. TEST PROCEDURES

(a) *Aerobic plate counts*

Comminute, blend, or stomach the various portions of meat composing the sample unit and withdraw the analytical unit. Combine the swabs to form an analytical unit, or treat each separately. Prepare dilutions and perform the APC test according to the directions given in ICMSF, 1978, p. 118. Spread plates are preferable to pour plates when examining raw meats.

The temperature used for incubation of plates should be related to the material being examined. To estimate numbers of mesophilic contaminants on freshly dressed carcasses it is advisable to incubate plates at 35 or 37°C. For meat that has been held under chilled conditions, plates should be incubated at 25°C to include psychrotrophs and thus give a better indication of the microbial growth that has occurred. Counts at 20°C give similar results after a longer incubation time.

(b) *Salmonella*

The test for salmonellae requires an analytical unit of 25 g and is based on positive or negative results using the method described in ICMSF, 1978, p. 163. For frozen products use a non-selective enrichment before selective enrichment and plating on selective agar media. Sample units should be taken from appropriate parts of carcasses, cuts of meat, and offals.

REFERENCES

Blaser, M.J. 1982. *Campylobacter jejuni* and food. *Food Technology*, *36* (3), 89–92

Codex.1981. *Codex Alimentarius Commission, Fourteenth Session, 1981: Report of the 17th Session of the Codex Committee on Food Hygiene, Washington, D.C. 17–21 November 1980*. ALINORM 81/13. Appendix II, General Principles for the Establishment and Application of Microbiological Criteria for Foods. Codex Alimentarius Commission, FAO, Rome

FAO/WHO (Food and Agricultural Organization / World Health Organization). 1979. *Microbiological Criteria for Foods*. Report of a Joint FAO/WHO Working Group on Microbiological Criteria for Foods, Geneva, 20–26 February 1979. (Document WG/Microbiol/79/1) WHO, Geneva

ICMSF. 1978. *Microorganisms in Foods. 1. Their Significance and Methods of Enumeration*, 2nd ed. University of Toronto Press, Toronto

– 1980. *Microbial Ecology of Foods*. Vol. 2. *Food Commodities*. Academic Press, New York

Roberts, T.A., MacFie, H.J.H., and Hudson, W.R. 1980. The effect of incubation temperature and the site of sampling on the assessment of the number of bacteria on red meat carcasses at commercial abattoirs. *Journal of Hygiene, Cambridge*, *85*, 371–380

12

Sampling plans for processed meats

Processed meats comprise a number of meat products, manufactured by many different technologies, including heat-processing, curing, drying, fermenting, acidifying, and packaging (ICMSF, 1980).

Processing changes the eating quality of meat. It also changes the composition of the microflora, and some processes kill or prevent growth of pathogens. Basically processing is intended to prolong the shelf-life of the products. The increase in shelf-life may range from a few days to several years. Processed meats with only a slight increase of shelf-life would not be candidates for international trade, while those given some degree of heat-treatment or curing combined with an adequate storage temperature move in international trade.

According to the FAO (1981), international trade during 1980 of processed meat and meat preparations amounted to about 1 million tonnes, of which sausages comprised about 115,000 tonnes. These FAO figures relate mainly to three groups of products: (a) shelf-stable cured canned meats; (b) perishable cured canned meats; and (c) cooked or uncooked, fermented, dried, or semi-dried sausages.

A. REASONS FOR SAMPLING

In the following section processed meats will be treated according to the outline in ICMSF, 1980, pp. 378–409. Discussion will deal with hazards (public health or spoilage), international trade of the products, and the possibility and feasibility of instituting means of control other than by microbiological criteria.

TABLE 20

Sampling plans and recommended microbiological limits for processed meats[a]

Product	Test	Method reference[b]	Case	Plan class	n	c	Limit per gram m	M
Dried blood, plasma, and gelatin	*Staph. aureus*	220	8	3	5	1	10^2	10^4
	C. perfringens	264	8	3	5	1	10^2	10^4
	Salmonella	163[c]	11	2	10	0	0	
'Roast' beef[d]	*Salmonella*	163[c]	12	2	20	0	0	
Pâté	*Salmonella*	163[c]	12	2	20	0	0	

a **Refer to text for appropriate applications.**
b Refers to page numbers in ICMSF, 1978.
c See also ISO 6579 (1981).
d Includes beef cooked in water baths; see Section B(f).

B. SAMPLING PLANS

(a) *Dried meats and dried animal products*

Dried meats like jerky, charqui, and biltong are mainly produced and consumed locally. Most dried meats found in international trade are intended for further processing. Further processing would normally consist of dry-blending with other food ingredients (e.g., dried soups) or rehydration and canning. In both cases the dried meats are subjected to conditions that either would not change or would reduce the hazards.

In international trade no evidence of hazard has been demonstrated, and no microbiological criteria are proposed.

Dried animal products such as dried whole blood, blood plasma, and gelatin are used as ingredients for formulated foods. When these ingredients are used in further processed products that are not subjected to adequate heat-treatment it is proposed to include sampling plans for *Staphylococcus aureus*, *Clostridium perfringens*, and *Salmonella* (Table 20).

For *Staph. aureus* and *C. perfringens* 3-class sampling plans, case 8 with $n = 5$, $c = 1$, $m = 10^2$, $M = 10^4$, are proposed. *Salmonella* should be considered in case 11 with a 2-class sampling plan with $n = 10$, $c = 0$.

(b) *Raw cured meats, high a_w*

High a_w meat products are defined as those with an a_w above 0.92, such as occurs in bacon produced by a Wiltshire or similar process (ICMSF, 1980, pp. 383–388).

In Wiltshire curing sides or parts of sides of pork, with or without bones, are injected with brine before either tank- or dry-curing. Smoke may be applied, but the product is not heated. There are many other types of manufacture of bacon, but this is the most common for bacon in international trade.

The microbiology of bacon is characterized by a rapid change from the initial gram-negative flora to gram-positive salt-tolerant micrococci and lactic acid bacteria. Only where curing is inadequate or the pH of the meat (usually the collar) is high will putrefactive spoilage dominate. Otherwise, spoilage is characterized by sliming.

Staphylococci could be considered to be of concern but are not able to compete with the normal saprophytic flora. Applying microbiological criteria at manufacture can make little contribution to the control of spoilage.

Salmonella is not a major concern because the product is cooked immediately before consumption.

(c) *Raw, cured, or salted meats and natural casings, low a_w*

Low a_w meat products are defined (ICMSF, 1980) as those having a water activity of 0.92 or below. This group includes whole cuts of meat (e.g., Bündener Fleisch, Parma ham, Westphalian ham, Smithfield ham, Prosciutti), non-fermented dried sausages, and natural casings, all of which are in international commerce.

Although some of the raw cured or salted meats have been implicated in outbreaks of food-borne illness, particularly botulism, the incidents have usually involved home-prepared products. Commercial products have had a very good record of safety. Growth of non-pathogenic bacteria occurs during the process of salting, curing, and drying of many of these products and can contribute to their flavour characteristics. Hence, microbiological criteria need not be applied to these commercial products. Cured, dried non-fermented sausages are in international trade but the only spoilage problem appears to be mould formation on the casing. No public hazards have been identified, and no microbiological criteria are proposed.

Natural casings from international commerce are generally packed in salt and have not been identified as a vehicle of food-borne illness. Survival and growth of some salt-tolerant bacteria occurs in salted casings, but *Salmonella* indigenous to the casings does not survive for more than two or three weeks after salting. Microbiological criteria are unnecessary after salting.

(d) *Fermented sausages*

Cured fermented sausages include both semi-dry and dry types. Both types undergo a fermentation process which should rapidly lower the product pH to <5.5. At this point semi-dry products may or may not be heated but if the pork used contains *Trichina* heating to 58.3°C may be mandatory. Other semi-dry sausages may only receive light heating (e.g., Thuringer). Dry sausage is placed in ''dry rooms'' with controlled humidity for an extended period after fermentation. During this ageing process, the product loses moisture and the a_w falls to c. 0.88, which is sufficiently low to make the heat-treatment to control *Trichina* unnecessary.

In both semi-dry and dry fermented sausages the main hazard is the growth of *Staph. aureus* during the initial stages of the fermentation before the pH of the mix falls sufficiently to inhibit its growth.

It has been suggested that microbiological criteria for *Staph. aureus* (not more than 100,000 per gram) be applied at the conclusion of the fermentation process or at the beginning of the drying process. More benefit could be obtained by a controlled acidulation to below pH 5.4 (using starter cultures or chemical acidulants such as glucono-delta-lactone) than by microbiological testing. *Salmonella* derived from the raw ingredients may survive processing but is unlikely to multiply. Total counts criteria are inappropriate because of growth of fermenting organisms.

The American Meat Institute has recently formulated Good Manufacturing Practices in connection with the production of fermented sausages (AMI, 1982). A maximum pH of 5.3 is prescribed in both dry and semi-dry fermented sausages.

(e) *Cured meats packed under vacuum*

Some of these products support the growth of *Clostridium botulinum* (Pivnick and Barnett, 1965) and *Salmonella* (Davidson and Webb, 1972) at abuse temperatures. The developing flora in vacuum-packaged cured meats is composed almost entirely of lactic acid bacteria. Large numbers of these organisms may develop without objectionable organoleptic changes. Despite the fact that these products are often abused through inadequate refrigeration, outbreaks of staphylococcal food-poisoning are seldom traced to them, probably because staphylococci fail to grow in competition with other microorganisms. Salmonellae are capable of survival during protracted storage of these products at 5°C and can grow when the meats are incubated at ambient temperature (Goepfert and Chung, 1970), but these products have not proved to be a significant *Salmonella* hazard.

It is assumed that sliced bacon will be heated prior to consumption and, in addition, that the growth of lactic acid bacteria will afford protection against growth of *Staph. aureus*. There is no evidence of food-borne disease from sliced bacon and thus criteria are not proposed.

Ham is sometimes involved in food-borne illness due to *Staph. aureus*, usually as a consequence of contamination during slicing or handling, followed by temperature abuse. Vacuum-packed sliced hams are rarely implicated. During storage of vacuum-packed meats, including hams, under refrigeration a large population of lactic acid bacteria develops in whose presence staphylococci fail to grow. There is no evidence that criteria at manufacture or at port of entry would reduce the hazard.

Frankfurters and bologna have also been implicated occasionally in food-borne illness, but control of the production process affords greater assurance of safety than would the application of criteria.

(f) *Cooked uncured meats*

Cooked uncured meats (e.g., roast-beef joints and cuts) are increasingly important in international trade. Their involvement in food-borne salmonellosis could be avoided by proper heat-processing, prevention of post-process contamination, rapid cooling, and subsequent appropriate time-temperature control in the home or food-service establishment. A criterion for *Salmonella* is proposed: case 12, $n = 20$, $c = 0$ (Table 20). There is no evidence that aerobic plate count (APC) limits would serve a useful purpose.

Considerable international trade exists in pasteurized comminuted meat products such as pâtés. At port of entry they frequently have high APCs and are occasionally associated with food-borne illness. Control is by adequate heat-processing, prevention of recontamination, and appropriate chill storage. Microbiological data are inadequate, but these products may contain *Salmonella*. If there is reason for concern, the appropriate sampling plan for *Salmonella* would be case 12, $n = 20$, $c = 0$ (Table 20).

(g) *Perishable cured canned meats*

The most important meat products of this kind in international trade are pasteurized canned cured hams and pork shoulders. A high proportion of these products are heated to centre temperatures of only 65 to 75°C. They should be transported under refrigeration, and should be labelled 'perishable, keep under refrigeration' or with words to similar effect. While the importance of testing such products is recognized, the high price of canned hams and similar foods is a serious obstacle to intensive examination. For these products, data on thermal processing, water supply, seam inspection, and chemical composition should be available, together with records of temperature during shipping. If all such data are satisfactory, no testing is necessary.

The Codex Alimentarius Committee for Processed Meat and Poultry Products is currently considering sampling plans for this category of products. Until the results become available the following is offered for guidance.

1. Sampling procedure at port of entry

If adequate production records are not available, sampling should be done according to the following scheme (see also Table 21).

a Measure the air temperature in the space between containers, preferably with an electronic measuring device.

b Take 10 containers at random from the lot. Identification of individual containers at this point is unnecessary. Examine the 10 containers for 'swells' and defects (see footnote, Table 21).

c If the temperature does not exceed 10°C or any specified lower temperature and if no swells or defective containers are found accept the lot.

If one or more defective or swollen containers are found, hold the lot and determine the cause by more extensive sampling (see Chapter 5, Section N, Investigational Sampling).

If the temperature exceeds 10°C or any specified lower temperature, or if there is reason to suspect that these temperatures have been exceeded during transport, proceed to step *d*.

TABLE 21

Sampling plans for perishable canned cured meats (to be used when processing data are not available or are unsatisfactory)

	Test	Case	Plan class	*n*	*c*	Limit per gram *m*	*M*	Acceptance
Step 1	Visual inspection for swells and defective containers*	–	2	10	0	–	–	Reject if 1 or more defective cans are found. Proceed to step 2 if no defective cans are found.
Step 2	Measurement of air temperature between cans	–	2	10	0	–	–	Accept if temperature is below 10°C. If higher proceed to step 3.
Step 3	Determination of aerobic plate count in centre and periphery	–	3	5	2	10^3	10^4	See Section G of this chapter for interpretation of the presence of micro-organisms.

* A defective container is a container which is of faulty manufacture, is improperly closed, or has been damaged in such a manner that may permit contamination of the contents following the heat-process (post-processing contamination).

d Take 5 containers from the warmer places in the lot and hold the lot. Proceed to step *e*.

e Identify the 5 containers selected under *d* and send them to a laboratory for microbiological examination. Transport should take place under refrigeration, at 10°C or less.

2. Laboratory analysis

a In the laboratory draw sample units from the 5 containers aseptically so as to obtain one unit from the centre and one from the periphery of each container.

b Examine for APC (ICMSF, 1978, p. 118). A direct Gram stain and an anaerobic count may sometimes be helpful. Rarely, the only bacteria present in large numbers are anaerobes.

c Reject if an analytical unit from one or more of the 5 containers has a viable count exceeding 10^4 per gram. Also reject if one or two analytical units from 3 or more containers show a viable count higher than 10^3 per gram. Otherwise, accept.

d When rejection is indicated an investigation into the cause of the problem might be appropriate and the supplier informed of the results. It should be emphasized that the values for *m* and *M* in this context do not necessarily reflect any spoilage or health hazard, but merely indicate degrees of undesirable commercial practice. The limits have been established from data derived from very large numbers of analyses over several years and are known to be attainable under GCP. It has been found that at export counts are normally less than 10^2 per gram. If the lot does not meet this criterion, the decision on the action to be taken should be one of the options given in Chapter 2.

(h) *Shelf-stable cured or uncured canned meats*

Shelf-stable meats have undergone a treatment sufficient to guarantee the subsequent safety and stability of the product at normal ambient temperatures. Canned cured and canned uncured meat products are considered with other low-acid canned foods in Chapter 26.

REFERENCES

AMI (American Meat Institute). 1982. *Good Manufacturing Practices: I. Voluntary Guidelines for the Production of Dry Fermented Sausages. II. Voluntary Guidelines for the Production of Semi-dry Fermented Sausages.* AMI, Washington, DC

Davidson, C.M., and Webb, G. 1972. The behaviour of salmonellae in vacuum-packaged cooked cured meat products. *Proceedings of the 18th Meeting of Meat Research Workers, Guelph, Canada.* University of Guelph, Guelph, Ontario

FAO (Food and Agriculture Organization) 1981. *Food Trade Yearbook*, Vol. 34. FAO Statistics Series No. 35, Rome

Goepfert, J.M., and Chung, K.C. 1970. Behaviour of salmonellae in sliced luncheon meats. *Applied Microbiology*, *19*, 190–192

ICMSF. 1978. *Microorganisms in Foods. 1. Their Significance and Methods of Enumeration*, 2nd ed. University of Toronto Press, Toronto

– 1980. *Microbial Ecology of Foods*. Vol. 2. *Food Commodities*. Academic Press, New York

ISO (International Organization for Standardization) 1981. *Microbiology – General Guidance on Methods for the Detection of Salmonella.* International Standard 6579. ISO, Geneva

Pivnick, H., and Barnett, H. 1965. Effect of salt and temperature on toxinogenesis by *Clostridium botulinum* in perishable cooked meats vacuum-packaged in air-impermeable plastic pouches. *Food Technology*, *19* (5), 140–143

13

Sampling plans for poultry and poultry products

There has been a steady increase in trade on the world poultry market. In 1980, over 1.4 million tonnes of poultry were exported from 45 poultry-producing countries (FAO, 1981). Imports by the industrialized and non-industrialized countries were nearly equal. Most of the product is frozen; however, there is some trade in fresh poultry between closely situated countries. There is also some trade in heat-processed and dried poultry products.

A. REASONS FOR SAMPLING

Raw poultry products are frequently contaminated by food-borne pathogens (e.g., *Salmonella*, *Staphylococcus aureus*, *Clostridium perfringens*, *Campylobacter fetus* subsp. *jejuni*, and *Yersinia enterocolitica*; ICMSF, 1980, pp. 419–50). Disease surveillance reports frequently identify poultry as vehicles in outbreaks of salmonellosis, staphylococcal food-poisoning, *C. perfringens* enteritis, and other enteric illnesses (Bryan, 1980; Hepner, 1980; Horwitz and Gangarosa, 1976; Todd, 1978; FAO/WHO, 1979). The prevalence of salmonellae in raw fresh and frozen poultry is of great concern and approaches 80% in some countries.

Contaminated chickens and turkeys bring salmonellae into kitchens and give rise to human salmonellosis, either directly or through cross-contamination to other foods (ICMSF, 1980, pp. 838–61; Bryan, 1981). Such outbreaks commonly occur from foods served in food-service establishments, hospitals, and other institutions. Deaths are infrequent, but they do occur in debilitated persons and infants.

Programs for decreasing *Salmonella* infection to low levels in fowl have been known for many decades; these are control of *Salmonella* in breeding flocks, prompt fumigation of eggs for hatching, rigid sanitary control in hatcheries, and provision of *Salmonella*-free feed. Where practised, these measures can greatly reduce the prevalence of *Salmonella*-contaminated carcasses, leading to low incidence of human infection from poultry. Most countries, however, have few or ineffective programs to combat the *Salmonella* problem, increased cost of poultry products being the most common reason given. An effective, but at present commercially unavailable, treatment could be the irradiation of packaged poultry products (ICMSF, 1980, pp. 455–457).

Spores of *C. perfringens* survive in some heat-processed poultry products. Non-spore-forming pathogens may survive if heat-processing is inadequate, and they may also survive curing, smoking, or drying processes. Poultry products are subject to post-processing contamination from staphylococci when sliced, wrapped, or otherwise handled by workers in processing plants and food-service establishments, and by persons who prepare foods in homes.

B. SAMPLING PLANS

(a) *Fresh and frozen raw poultry*

At the time of preparing the previous edition of this book, it was thought that the sampling plan for *Salmonella* in poultry of case 10, $n = 5, c = 0, m = 0$ was not feasible; hence the c value was set at 1 (ICMSF, 1978b, pp. 137–139). Even this has proved to be unrealistic under current conditions of intensive poultry production and processing in most countries (FAO/WHO, 1979; ICMSF, 1980, pp. 438–442).

The proportion of salmonellae-contaminated poultry carcasses is determined mainly by the quantity of incoming infected or contaminated live fowl and by subsequent spread during defeathering and other processing (ICMSF, 1980, pp. 410–50). Unfortunately, even good manufacturing practices in a plant do not appreciably affect the situation (FAO/WHO, 1979). Until changes are made to prevent or minimize the infection and contamination of fowl on farms, or until a method of decontamination (e.g., irradiation of packaged carcasses or treatment with an acceptable chemical) is routinely applied, it is questionable whether practicable port-of-entry microbiological criteria can be set for *Salmonella* on frozen raw poultry without risk of eliminating poultry as a food. Educational programs are needed to maintain food safety in kitchens. Training must emphasize thorough cooking, prevention of cross-contamination, rapid cooling of cooked products that are not to be eaten immediately, and thorough reheating, particularly for poultry being prepared in medical care and geriatric institutions.

Investigational sampling plans (see Chapter 5, Section N) may be used to (1) determine the extent of contamination of a feed supply or a product with salmonellae; (2) identify or monitor critical control points of hatchery, farm, or processing operations; (3) trace a source of contamination; or (4) evaluate effectiveness of control measures. Such data may help to convince producers and others that a serious problem exists and that steps should be taken towards the elimination of *Salmonella* in live fowl, thereby controlling the problem at source. Initially, a lenient sampling plan for *Salmonella* will be needed (e.g., $n = 5$, $c = 3$, $m = 0$); but as control measures (such as the use of *Salmonella*-free chicks and feeds) are initiated and maintained, the plan should be made more stringent by reducing c.

Although *Campylobacter fetus* subsp. *jejuni* and *Y. enterocolitica* are frequent contaminants of raw poultry products (deBoer et al., 1982; Leistner et al., 1975; Norberg, 1981; Simmons and Gibbs, 1979), and poultry-associated outbreaks of campylobacteriosis have been reported (Cunningham, 1982), criteria are not considered necessary. Thorough cooking, care to avoid cross-contamination, and proper storage of cooked products will minimize the risks associated with these bacteria.

Clostridium perfringens is a part of the normal intestinal flora of poultry and is shed in their faeces, and is widespread in their environment. Its presence on poultry carcasses cannot be avoided, but it will not multiply on chilled carcasses. While cooking destroys vegetative cells of *C. perfringens*, spores survive and may even be activated. Also the redox potential of carcasses is lowered by cooking. Temperature abuse of cooked poultry can promote spore germination and multiplication of the resulting vegetative cells. A criterion for *C. perfringens* in or on raw poultry for port-of-entry sampling is not recommended. Prevention of illness must therefore be based on time-temperature control after cooking.

Staphylococcus aureus is a part of the normal flora of poultry, and even adherence to good processing practices cannot totally prevent carcass contamination. However, staphylococci do not compete well with normal spoilage flora of poultry and are not likely to multiply on chilled carcasses. They become a problem only if cooked poultry is contaminated and subsequently subjected to time-temperature abuse. A criterion for *Staph. aureus* in raw poultry is not recommended for port-of-entry sampling. Poultry eaten shortly after cooking or not handled after cooking and rapidly cooled is not associated with staphylococcal food-poisoning.

A sampling plan for *Staph. aureus* is useful to evaluate carcass contamination from picking (plucking) machines in which staphylococci can accumulate particularly and contaminate carcasses passing through. A sampling plan of case 4 (staphylococci as an indicator of a source of contamination), $n = 5$, $c = 3$, $m = 10^3$, $M = 10^4$ per cm^2 of swabbed surface might be initially considered for in-plant quality control.

It is also considered unnecessary to have microbiological criteria for other indicator organisms (Enterobacteriaceae, coliforms, faecal coliforms, *Escherichia coli*, and enterococci) because they are part of the normal intestinal flora of poultry. Some will be spread among carcasses even under good processing practices (FAO/WHO, 1979). Some strains of species that fall into the indicator organisms categories (such as certain coliforms and *Enterobacter*) are psychrotrophic and can multiply on refrigerated raw poultry carcasses and products.

Microorganisms responsible for spoilage of poultry multiply during refrigeration, even at recommended temperatures of 1 to 4°C (ICMSF, 1980, p. 443). A commonly recommended microbiological method for estimating the shelf-life of chilled poultry, an aerobic plate count (APC) at 0 to 5°C, is impracticable for use at port-of-entry because it requires incubation for 10 to 14 days.

In the previous edition of this book, a criterion for APC (20°C) of case 1, $n = 5$, $c = 3$, $m = 5 \times 10^5$, $M = 10^7$, was suggested (ICMSF, 1978b, p. 138). This criterion is achievable for processed chickens, but not necessarily turkeys, in modern processing plants that follow good processing practices (CEC, 1979) and is a useful guideline. It is not recommended, however, as a port-of-entry microbiological criterion or as a basis for estimating shelf-life.

Irradiation could eliminate *Salmonella* from processed raw and frozen poultry (ICMSF, 1980, p. 455). A criterion of case 10, $n = 5$, $c = 0$, $m = 0$, would be applicable to *Salmonella* in raw or frozen irradiated poultry. It is premature to propose other microbiological criteria for irradiated poultry before these products enter international trade, and until additional microbiological data and experience become available.

(b) *Frozen, heat-processed poultry products*

There is increasing international trade in heat-processed poultry products (i.e., turkey rolls and entrée items such as fried chicken, chicken à la king, and pot pies). (See Chapter 20, Cereals and Cereal Products, for suggested microbiological criteria and further discussion of pot pies

TABLE 22

Sampling plans and recommended microbiological limits for poultry and poultry products

Product	Test	Method reference[a]	Case	Plan class	n	c	Limit per gram m	Limit per gram M
Cooked poultry meat, frozen; to be reheated before eating (e.g., prepared frozen meals)	*Staph. aureus*[b]	218	8	3	5	1	10^3	10^4
	Salmonella	163	10	2	5	0	0	
Cooked poultry meat, frozen, ready-to-eat (e.g., turkey rolls)	*Staph. aureus*[b]	218	8[c]	3	5	1	10^3	10^4
			9[c]	3	10	1	10^3	10^4
	Salmonella	163	11	2	10	0	0	
Cured and/or smoked poultry meat	*Staph. aureus*[b]	218	9	3	10	1	10^3	10^4
	Salmonella	163	11	2	10	0	0	
Dehydrated poultry products	*Salmonella*	163	11	2	10	0	0	
Raw chicken (fresh or frozen), during processing	APC	118 (20°C)	1	3	5	3	5×10^5	10^7[d]

a Refers to page numbers in ICMSF 1978a, where the methods are described. Use analytical unit sizes recommended in the methods, except where otherwise indicated.
b If either packaged or repackaged after processing; not for products processed in packages that are kept closed until time of final preparation.
c Case depends on whether (case 9) or not (case 8) subsequent time-temperature abuse of the products is likely.
d **Not for port-of-entry sampling**; for use to evaluate production and operations and critical control points in processing plants only.

and other blended products with pastry shells or covers.) Presence of *Salmonella* may be due to failure of heat processes or cross-contamination from the raw to the cooked product (ICMSF, 1980, pp. 850–859). A microbiological criterion for *Salmonella* in heat-processed poultry products is given in Table 22.

Staphylococci and faecal indicators are useful monitors of contamination of frozen cooked poultry products. Such monitoring is unnecessary when products are adequately cooked in plastic casings or bags, but it is essential when these products are handled, cut, repackaged after heat-processing, or otherwise subjected to any operation that may permit contamination. Although the data base is limited, microbiological criteria for *Staph. aureus* are recommended. Case 8 applies, because enterotoxins of *Staph. aureus* withstand the time-temperature exposure of reheating. If there is likelihood of time-temperature abuse, case 9 would apply (see Table 22). Staphylococci survive freezing better than the Gram-negative organisms (such as *E. coli*) and can be a useful indicator of post-heat-process contamination.

(c) *Smoked, cured, and dried poultry products*

Although there is limited international trade in smoked, cured, and dried poultry products other than dried eggs, and the need for microbiological criteria has not clearly been established, these products should be free from *Salmonella*. The presence of *Salmonella* constitutes a health hazard and case 11 applies (Table 22). Dried products are shelf-stable, as long as packages are intact, until rehydrated.

Monitoring for APC, *E. coli*, and *Staph. aureus* would give some indication of inadequate processing (heating, curing, smoking), post-process contamination, or storage at temperatures above those recommended for frozen or refrigerated distribution and storage. These products are likely to become contaminated during slicing or wrapping. The a_w level may inhibit common spoilage organisms but not necessarily *Staph. aureus*.

Because only limited microbiological data on these commodities are available, criteria for APC or *E. coli* are not recommended. However, a criterion is suggested for *Staph. aureus* on smoked or cured poultry products (Table 22). Case 9 is applicable because these products are frequently distributed or stored at temperatures conducive to multiplication of staphylococci.

C. SAMPLING PROCEDURES

Take 5 or 10 sample units, as appropriate, from each lot. Sample units of fresh or frozen raw poultry should be taken from more than one carton where possible. Sterile disposable plastic gloves are useful for holding carcasses. Use sterile instruments to cut portions, if necessary, and to collect sample units (usually 200 g) of comminuted, cooked, dehydrated, smoked, or cured poultry, and aseptically transfer them to sterile containers. (See Chapter 9, Collecting and Handling Sample and Analytical Units, for more information on aseptic sampling, identifying samples, and transporting samples to the laboratory.)

Use the shake and rinse procedure for sampling fresh or thawed raw poultry carcasses to be tested for *Salmonella*. In this procedure, the carcass is placed aseptically into a heavy-duty, plastic bag (which has not been previously used) containing 300 ml buffered peptone water. The bag is closed, and the enclosed carcass shaken vigorously, vertically and horizontally, for 30 seconds. The entire surface and cavity of the carcass should come into contact with the water during the shaking. Pour the rinse into a sterile container and keep it at 4°C until it is used in the laboratory. If raw turkey carcasses are to be tested, use 500 ml buffered peptone water. This procedure is difficult to use for large turkeys because they are heavy and it is difficult to shake them effectively. Swabbing is a suitable alternative method (see below).

The shake and rinse samples are not usually practicable for 'on-the-line' sampling. In such situations, rub a swab (which has been moistened in sterile peptone water) over the skin and, if practicable, internal regions of the carcass. Include both neck and vent in the swabbed areas. Alternatively aseptically cut portions of neck skin and take interior scrapings. When these procedures are used, however, the likelihood of detecting salmonellae may be less than with the shake and rinse procedure.

The recommended procedure for sampling raw carcasses for APC is to collect samples of neck skin. Use sterile instruments to grasp the carcass and to cut sample units from carcasses and aseptically put them into a sterile container.

D. TEST PROCEDURES

Carcass-rinse and swab sample units become the analytical units for the *Salmonella* or other test. Thaw ('drip') water from raw poultry can also be used as analytical units. Rinse and thaw waters can also be used for tests for other organisms. Incubate each as prescribed in ICMSF, 1978a, p. 163. When testing skin sample units for *Salmonella*, put cut portions of skin into pre-enrichment broth according to procedures in ICMSF, 1978a, p. 164. For APC, weigh 20–50 g of neck skins and homogenize this analytical unit with 0.1% peptone saline (ratio 1:9) in a stomacher or blender.

Sample units from frozen poultry should be thawed at 1 to 5°C, preferably within 18 hours. Comminute and blend a sample unit, and from this composite weigh out analytical units for the APC, *Staph. aureus*, and other counts and for the *Salmonella* test according to directions given in ICMSF, 1978a.

REFERENCES

Bryan, F.L. 1980. Foodborne diseases in the United States associated with meat and poultry. *Journal of Food Protection, 43*, 140–150

– 1981. Current trends of salmonellosis in Canada and the United States. *Journal of Food Protection, 44*, 394–402

CEC (Commission on the European Communities). 1979. *Microbiological Methods for Control of Poultry Meat*, Study P203, VI/5021/79-EN, EEC, Brussels

Cunningham, F.E. 1982. Microbiological aspects of poultry and poultry products – An update. *Journal of Food Protection, 45*, 1149–1164

deBoer, E., Hartog, B.J., and Oosterom, J. 1982. Occurrence of *Yersinia enterocolitica* in poultry products. *Journal of Food Protection, 45*, 322–325

FAO. 1981. *1980 FAO Trade Yearbook*. Vol. 34. FAO, Rome

FAO/WHO. 1979. *Microbiological Criteria for Foods*. Report of a Joint FAO/WHO Working Group on Microbiological Criteria for Foods, Geneva, 20–26 February, 1979 (Document WG/Microbiol/79/1). WHO, Geneva

Hepner, E. 1980. Food poisoning and *Salmonella* infection in England and Wales, 1976–1978. An analysis of reports to the Communicable Disease Surveillance Centre of the Public Health Laboratory Service. *Public Health, 94*, 337–349

Horwitz, M.A., and Gangarosa, E.J. 1976. Foodborne disease outbreaks traced to poultry, United States, 1966–1974. *Journal of Milk and Food Technology, 39*, 859–863

ICMSF. 1978a. *Microorganisms in Foods. 1. Their Significance and Methods of Enumeration*, 2nd ed. University of Toronto Press, Toronto

– 1978b. *Microorganisms in Foods. 2. Sampling for Microbiological Analysis: Principles and Specific Applications*. University of Toronto Press, Toronto

– 1980. *Microbial Ecology of Foods*. Vol. 2. *Food Commodities*. Academic Press, New York

Leistner, L., Hechelmann, H., Kashiwazaki, M., and Albertz, R. 1975. Nachweis von *Yersinia enterocolitica* in Faeces and Fleisch von Schweinen, Rindern und Geflugel. *Fleischwirtschaft, 11*, 1599–1602

Norberg, P. 1981. Enteropathogenic bacteria in frozen chicken. *Applied and Environmental Micobiology, 42*, 32–34

Simmons, N.A., and Gibbs, J.F. 1979. *Campylobacter* spp. in oven-ready poultry. *Journal of Infection, 1*, 159–162

Todd, E.C.D. 1978. Foodborne disease in six countries – a comparison. *Journal of Food Protection, 41*, 559–565

14

Sampling plans for feeds of animal origin and pet foods

Large quantities of feeds destined for animals and increasing quantities of dehydrated pet foods are in international trade. Feeds and pet foods have constituents of vegetable and animal origin, both of which can be contaminated with many kinds of microorganisms or their metabolites, including those of animal or human health significance. Although the absence of other pathogens in feeds cannot be guaranteed, the main organisms of concern are salmonellae. Fungal metabolites are potentially hazardous to animals and to human health through residues in foods of animal origin (ICMSF, 1978).

A. FEEDS

Animals ingesting feeds contaminated with *Salmonella* may contract clinical disease or subclinical infections and excrete salmonellae for long periods. This fact has become important with the current increase in world trade of feeds, and fish meal and meat-bone meal have been responsible for the spread of certain *Salmonella* serotypes, such as *Salmonella agona*, world-wide (Clark et al., 1973).

Animals excreting salmonellae may endanger public health either by direct contact (e.g., turtles in the home) or indirectly when foods of animal origin are contaminated as in slaughtering or milking. For discussions of factors in addition to contaminated feeds, such as the environment, see ICMSF, 1980, pp. 459–469; Riemann and Bryan, 1979; WHO, 1981; Silliker, 1980.

It is accepted that elimination of *Salmonella* from feeds will not guarantee the absence of *Salmonella* in animals. However, to assist in the reduction of the hazard, efforts should be made to reduce the contamination of animal feeds either by producing *Salmonella*-free feed components of animal and vegetable origin or by subjecting the compounded feed to a decontaminating treatment like certain methods of pelleting, heating, or irradiation (Statutory Instruments, 1981a, b; van Schothorst and Brooymans, 1982). However, criteria for livestock feed have not been proposed because they are unlikely to reduce the prevalence of *Salmonella* in animals effectively unless other measures are taken at the same time. Feed subjected to heating, irradiation, and some pelleting processes should be free from *Salmonella*. For such feeds a sampling plan $n = 10$, $c = 1$, or more desirably $c = 0$, is recommended. Feeds subjected to other pelleting processes, not primarily intended to destroy salmonellae, cannot be expected to meet this criterion.

Heating during the fat-extraction stage of production of protein-rich feed components of animal origin (e.g., fish meal) or of vegetable origin (e.g., soya flour, cotton-seed meal) yields products free from *Salmonella* and other heat-sensitive contaminants. However, post-process contamination frequently occurs, sometimes reintroducing *Salmonella* by cross-contamination with raw material and from common machinery as well as aerosols from grinders, dust, flies, and excrement of birds and rodents.

During processing good hygienic housekeeping, keeping dry areas of the factory dry, completely separating raw material and finished product, and preventing recontamination are the most important measures to ensure *Salmonella*-free products. Environment and line sampling are more effective in *Salmonella* control than testing end-products (ICMSF, 1980, pp. 459–469; van Schothorst and Oosterom, 1984). At port of entry, testing for the presence of *Salmonella* may identify contaminated consignments.

B. PET FOODS

Dehydrated, extruded, pelleted, and intermediate-moisture pet foods can be produced free from *Salmonella* even though some of the raw materials contain it. In contrast pet foods produced by blending such raw ingredients cannot be produced free from *Salmonella* because no bactericidal treatment is applied (ICMSF, 1980; van Schothorst and Brooymans, 1982). Irradiation might be considered in this case (WHO, 1981).

Canned pet foods are not dealt with in this chapter because methods for control do not differ in principal from those described in Chapter 26 (Shelf-Stable Canned Foods).

TABLE 23

Sampling plans and recommended microbiological limits for pet foods[a]

Product	Test	Method reference[b]	Case	Plan	n	c	m	M
Intermediate moisture pet foods	*Salmonella*	163–172	11	2	10	0	0	–
Dry pet foods not to be reconstituted	*Salmonella*	163–172	11	2	10	0	0	–
Dehydrated pet food to be reconstituted	*Salmonella*	163–172	12	2	20	0	0	–

a Case 11 could also be used for monitoring animal feeds for *Salmonella* (see this chapter).
b Refers to page numbers in ICMSF, 1978; see also ISO 6579 (1981).

Frozen raw meat of origins not covered in Chapter 11 (e.g., horse and kangaroo meat) and meat offals are shipped as pet food to many countries of the world. These pet foods may contain *Salmonella* and may be a source of contamination of the kitchen as well as sources of infection for animals. These pet foods are occasionally eaten by humans, though this may be regarded as misuse of the product. Taking into account the present status of the *Salmonella* problem (FAO/WHO, 1979), criteria for these pet foods at the port of entry are not warranted (WHO, 1981; cf. Chapter 11, Raw Meats; Chapter 13, Poultry). However, monitoring of product for investigational purposes would provide information on the magnitude of possible problems.

C. CHOICE OF MICROBIOLOGICAL CRITERIA

If the same considerations are applicable to the selection of cases for *Salmonella* in animal feeds as for human foods, case 11 and case 12 would be chosen for the various categories of feeds or pet foods. This choice is reasonable because the ranges of minimal infective doses which provoke clinical disease in humans and in animals are similar.

Case 11 was previously chosen for animal feeds and dry and intermediate-moisture pet foods, because their low levels of water activity prevent multiplication of salmonellae (ICMSF, 1980, p. 469). For dehydrated pet foods which are reconstituted and left at room temperature, case 12 applies (ICMSF, 1980, p. 469) (Table 23). Feed components of animal origin are not included.

The criterion for *Salmonella* in animal feeds should therefore, in principle, be case 11, $n = 10$, $c = 0$, $m = 0$. However, many feeds in international trade cannot meet this criterion. Decontamination (e.g., by irradiation, heating, or pelleting) would increase costs; and adequate facilities for these processes are not yet widely available. Therefore, it is recommended that consignments be monitored for the presence of *Salmonella* for epidemiological or investigational purposes.

In Chapter 9, guidance on sample collection, preparation, and analysis is given. The method of ICMSF, 1978, pp. 163–172, is recommended. It is, in principle, in agreement with ISO Standard No. 6579.

REFERENCES

Clark, G. McC., Kaufmann, A.F., Gangarosa, E.J., and Thompson, M. 1973. Epidemiology of an international outbreak of *Salmonella agona*. *Lancet*, ii, 490–493

FAO/WHO (Food and Agricultural Organization / World Health Organization). 1979. *Microbiological Criteria for Foods*. Report of a Joint FAO/WHO Working Group on Microbiological Criteria for Foods, Geneva, 20–26 February 1979. (Document WG/Microbiol/79/1) WHO, Geneva

ICMSF. 1978. *Microorganisms in Foods. 1. Their Significance and Methods of Enumeration*, 2nd ed. University of Toronto Press, Toronto

– 1980. *Microbial Ecology of Foods*. Vol. 2. *Food Commodities*. Academic Press, New York

ISO (International Organization for Standardization). 1981. *Microbiology – General Guidance on Methods for the Detection of Salmonella*. International Standard 6579. ISO, Geneva

Riemann, H., and Bryan, F.L. (eds). 1979. *Food-Borne Infections and Intoxications*. Academic Press, New York

Silliker, J.H. 1980. Status of *Salmonella* – Ten years later. *Journal of Food Protection, 43*, 307–313

Statutory Instruments.1981a. *The Diseases of Animals (Protein Processing) Order 1981*. Statutory Instruments No. 676. HMSO, London

– 1981b. *The Importation of Processed Animal Proteins Order 1981*. Statutory Instruments No. 677. HMSO, London

van Schothorst, M. and Brooymans, A.W.M. 1982. Effect of processing on microbial contaminants in feeds. In *Handbook of Nutritive Value of Processed Food*. Vol. 2. *Animal Feedstuffs*, ed. M. Rechcigl, Jr, 371–385. CRC Press, Inc., Boca Raton, USA

van Schothorst, M. and Oosterom, H. 1984. Enterobacteriaceae as indicators of Good Manufacturing Practices in rendering plants. *Antonie van Leeuwenhoek*, *50*, 1–6

WHO (World Health Organization). 1981. *Report of the WHO/WAVFH Round Table Conference on the Present Status of the Salmonella Problem (Prevention and Control)*. Bilthoven, The Netherlands, 6–10 October 1980. (Document VPH/81.27) WHO, Geneva

15

Sampling plans for milk and milk products

For purposes of applying microbiological criteria, dairy products may be divided into two broad groups: (a) the more perishable ('fresh') products such as milk, cream, flavoured milk and skim milk drinks, fresh cheese (cottage cheese), and fermented milks; and (b) the relatively stable products having extended shelf-life under appropriate conditions of storage, such as hard cheese, butter, dried milk products, ice cream mixes, evaporated (canned) milk, and sterilized or ultra-high-temperature (UHT) milk (for fluid consumption).

A. RELATIVELY PERISHABLE PRODUCTS

It is impracticable to apply end-product microbiological criteria to detect defective products in this category, which will have been widely distributed and probably consumed before microbiological examinations have been completed. However, periodic analyses (see below) are commonly made by local authorities to decide whether or not to continue to accept a product. Judgment is based on the performance record of the producer or processor over a period of time, such as the 3 out of 5 rate of compliance with the standard as specified in the U.S. *Grade A Pasteurized Milk Ordinance* (USPHS, FDA, 1978).

Health problems to be anticipated from pasteurized fluid milk and related products are likely to be of local concern, and control is normally in the hands of public health authorities. Pasteurized fluid milk rarely crosses frontiers, so agreements between exporting and importing countries appear as the best solution to the problem of control. Therefore, microbiological criteria are not proposed for pasteurized milk in international trade. The report of the joint FAO/WHO Expert Committee on Milk Hygiene (FAO/WHO, 1970) and ICMSF (1980b, pp. 482–486) provide useful information on milk control practices and disease hazards. Hygienic codes and methods of sampling and analysis are described in publications of the International Dairy Federation (IDF, 1973, 1980b, 1980c) and the American Public Health Association (APHA, 1978), and in the Grade A Pasteurized Milk Ordinance (USPHS, FDA, 1978).

The microbiological tests most often used are the aerobic plate count (APC) and the coliform test. In addition, the phosphatase test is used world-wide. If this enzyme is detected in allegedly pasteurized milk, it is almost certain either that the milk was underpasteurized or that raw milk had somehow entered the product.

B. RELATIVELY STABLE PRODUCTS

Milk products of this group are usually subject to test before being distributed and consumed. Those that are currently common in international trade and are of concern because of microbiological hazards they may present are dried milk and ripened cheese. While others may be important in international trade, either they are unlikely to present a microbiological health hazard or their safety and quality are not amenable to measurement by end-product microbiological criteria. For example, butter is an important commodity in international trade, but modern methods of manufacture ensure a relatively long shelf-life under normal commercial practice, and only rarely has illness been attributed to butter (see ICMSF, 1980b, pp. 775–776). The safety of UHT milk, as of other low-acid canned foods (including evaporated milk), depends primarily upon ensuring that the heat-processing is adequate and that container integrity is maintained. Routine sampling and testing of end-product will not provide adequate assurance of safety (see Chapter 10 and ICMSF, 1980a, pp. 32–36). However, it is recommended that industry monitor microbiologically, wherever applicable, critical control points in processing operations. To some extent ice-cream mixes occur in international trade. Ice cream itself is generally subject to control by local authorities. If such controls do not exist it is recommended that the WHO proposals (Christian, 1983, pp. 14–16) be followed for monitoring purposes.

(a) *Dried Milk*

Several types and grades of dried milk occur in commerce. Few outbreaks of food-borne disease due to dried milk have been reported in recent years. This fact undoubtedly reflects the effectiveness of (a) the extensive *Salmonella* testing of dried milk and subsequent action to prevent distribution of contaminated product and (b) the critical control measures now commonly practised in the industry (see ICMSF, 1980b, pp. 492–494; and Chapter 7, Control at Source) introduced following the outbreaks of staphylococcal food-poisoning and salmonellosis that spanned the years 1953–56 and 1964–65, respectively (Anderson and Stone, 1955; Armijo et al., 1957; Collins et al., 1968). Nevertheless, considering its intended use, dried milk is a sensitive product and may present serious health hazards if not manufactured and used according to good hygienic practices.

The various cases for concern are given in Tables 6 and 10 (pp. 43 and 74). The m and M values are given in Table 24. Case 2 for APC and case 5 refer to indicator measurements and are recommended in association with 3-class plans (see Table 10). Coliform organisms in dried milk may die out during storage. They should, nevertheless, be absent or present only in low numbers in dried milk, even when the product is freshly made. Thus, control of coliforms is necessary but their absence is not sufficient to qualify a product as acceptable. Ideally the m value for coliforms should be low (c. 1 per g) and, indeed, an m value of <3 was suggested in the previous edition of this book. Since the current method (3-tube MPN) is imprecise in estimating such a low number, a different value is indicated. It was desired to change this previous plan to one having essentially the same consumer's and producer's risks but using a higher m value. The adjustment can best be achieved by decreasing c from 2 to 1 but increasing m to 10. n remains at 5 and M at 100 (see Table 24). It is anticipated that this case 5 sampling plan for coliforms in dried milk will provide essentially the same assurance as the plan previously proposed ($n = 5$, $c = 2$, $m = <3$, $M = 100$).

Enterotoxigenic *Staphylococcus aureus* is a potential hazard in dried milk and should be absent or be present only in very low numbers. The hazard of enterotoxin being present in the dried product is likely only when deviations from Good Manufacturing Practice (GMP) provide opportunities for sufficient growth of the organism either in the milk before condensing or in the condensed milk prior to drying. The likelihood that conditions would permit such extensive growth is remote in modern dry-milk manufacture because of industry's awareness of the problem and of the measures taken to prevent it. There are, however, situations where the product might become contaminated after drying. Thus, if reconstituted milk is exposed to times and temperatures

TABLE 24

Sampling plans and recommended microbiological limits for dried milk and cheese

Product	Test[d]	Method reference[a]	Case	Plan	n	c	Limit per gram m	Limit per gram M
Dried milk	APC	118	2	3	5	2	3×10^4	3×10^5
	Coliforms	126–131	5	3	5	1	10	10^2
	Salmonella,[e] normal routine	163	10	2	5	0	0	
			11	2	10	0	0	
			12	2	20	0	0	
	Salmonella,[e] for high-risk population	163	10	2	15	0	0	–
			11	2	30	0	0	–
			12	2	60	0	0	–
Cheese, 'hard' and 'semi-soft' types[b, c]	*Staph. aureus*	220	8	2[f]	5	0	10^4	–

a Refers to page number in ICMSF, 1978, where the method is described. Use sample unit sizes recommended in the methods, except where otherwise indicated.

b In addition, such cheese should be aged for 60 days at not less than 4.4°C (40°F) when made from unpasteurized or unheated milk or curd (see Section B(b) 1).

c See also discussion on thermonuclease, Section B(b) 3.

d APC, see also IDF, 1970; coliforms, see also IDF, 1971; *Salmonella*, see also ISO, 1981.

e The case for *Salmonella* is to be chosen in accordance with whether the expected use of the product will reduce, cause no change in, or increase concern (see Table 10, p. 74)

f See Section B(b) 3 for discussion of sampling plan.

sufficient to permit growth of staphylococci – for example, temperature abuse during mass feeding operations in hospitals, schools, institutions, or in relief or emergency conditions – the hazard of enterotoxin could exist. For discussion of sampling plans for products used in this manner, see Chapter 24, Formulated Foods.

The presence of *Salmonella* in dried milk is a serious hazard. While current GMPs are effective in excluding *Salmonella* from dried milk, contaminated lots are occasionally found (see ICMSF, 1980b, p. 494). End-product testing for *Salmonella* is widely practised and is an important factor in identifying contaminated lots and in preventing them from entering trade. For routine sampling, in accordance with expected use of the product cases 10, 11, and 12 would apply (see Table 10, p. 74). Where the intended use is not known case 12 also would apply. More stringent sampling plans for cases 10, 11, and 12 are indicated when the product is intended for consumer groups of increased susceptibility, such as infants, the aged, and infirm (see Table 8, p. 55; Chapter 6, Section B; and Table 12, p. 78). The 2-class plans which are recommended are given in Table 24.

Recently, the Codex Committee on Food Hygiene proposed end-product specifications to be included in a Code of Hygienic Practice for Dried Milk (Codex 1981, 1983). Their proposed criteria are intended to apply only when the product is used for other than high-risk populations and are similar to those shown in Table 24 except that only one specification for *Salmonella* ($n = 15$, $c = 0$, $m = 0$) is recommended regardless of product use.

Microbiological guidelines were not recommended for inclusion in the Code of Hygienic Practice for Dried Milk. It was concluded that guidelines should be formulated by manufacturers taking into account the need to meet limits prescribed in end-product criteria. ICMSF believes this conclusion was prudent. It leaves the mechanism of compliance with the code flexible and at the option of producers or manufacturers.

Foods prepared especially for infants and children may consist wholly or in part of dried milks. These products are included in the Codex 'Code of Hygienic Practices for Foods for Infants and Children' (Codex, 1979) and are considered in Chapter 24 (Formulated Foods).

(b) *Cheese*

Throughout the world diverse procedures are used to produce various types of cheese from milk, which may be raw, pasteurized (phosphatase-negative), or heat-treated. For many types of cheese data on their bacteriological quality, from the hygienic point of view, are few and often have not been related to GMP. More data are necessary before meaningful bacteriological limits can be recommended for every cheese variety. Nevertheless, for some cheeses common in international trade, present knowledge and experience are sufficient to warrant proposals for microbiological criteria.

Although cheese has been the vehicle in outbreaks of food-borne illness, epidemiological evidence places it among the relatively infrequent vehicles of transmission.

1. Aged cheese made from raw milk

Many varieties of cheese are made from milk which has not been heated before the cheese-making process begins. Certain types, however, require heat-treatment of the curd at some point during the cheese-making process, such as during cooking.

Several pathogenic bacteria, including *Salmonella*, *Campylobacter*, and *Listeria*, can survive in cheeses made from raw milk. Some protection against the transmission of infectious disease, especially brucellosis, is achieved by requiring cheese made from other than pasteurized milk to be aged for at least 60 days at a temperature of not less than 4.4°C.

2. Aged cheese made from milk heated during the cheese-making process

Some types of cheese (e.g., Gruyere, Emmental, and Grana) are made from raw milk, but the curd undergoes a heat-treatment at approximately 53–55°C for about 1 hour during the cheese-making process. These heating processes reduce the bacteriological risk in such cheese.

3. Aged cheese made from pasteurized or 'heat-treated' milk

Cheese is made also from pasteurized milk (phosphatase-negative) or milk heated to a degree which may not provide a negative phosphatase test but does substantially reduce the number of microorganisms. The latter process is sometimes referred to as 'heat-treating' or 'thermization' of milk; the heat-treatment approximates 65.5°C for 15 to 20 seconds. Such heat-treatments destroy or greatly reduce the number of heat-labile organisms, including staphylococci and other pathogens (Zottola and Jezeski, 1969; Zottola et al., 1969). The presence of excessive numbers of such organisms in the final product results either from contamination of the milk after heating and subsequent growth or from some faulty practice or accident during the cheese-making process (e.g., starter failure).

The most serious health risk presented by cheese shipped in international trade is from staphylococcal enterotoxin. Available information about staphylococci in cheese and factors influencing their growth and production of enterotoxins is limited primarily to hard (Cheddar and similar types, and Swiss) and semi-soft (Roquefort, Blue, Gouda, Brick, and Mozzarella) varieties (ICMSF, 1980b, pp. 508–510; Devoyod et al., 1968, 1969). The maximum growth of staphylococci occurs when whey drainage is complete and generally prior to salting or brining. Later, during the ageing period, they die slowly and only a relatively small proportion of the earlier numbers will be recovered when cultured. However, any enterotoxins produced will remain.

Growth of enterotoxigenic staphylococci to c. 1–5 million per gram of cheese during cheese-making is necessary to produce amounts of enterotoxin detectable by current assay methods (Tatini et al., 1971a, 1971b, 1973). Heat-treatments less severe than those indicated above (65.5°C for 15 to 20 seconds), when applied to milk before or after the cheese-making process begins, may leave a residual population of coagulase-positive staphylococci. In addition, insanitary equipment may recontaminate the milk after heat-treatment. However, the most important factor permitting the growth of staphylococci, and of many other undesirable bacteria, during cheese-making is failure of the lactic acid starter culture to develop normally ('starter failure'). This usually results from infection of the starter culture by bacteriophage, or from the presence of small amounts of antibiotics or other inhibitory substances in the milk. Staphylococci found in ripened cheese probably represent only a fraction of the population present in the unripened cheese. Thus, the reliability of criteria for staphylococci in ripened cheese is limited, and this fact must be taken into account when test results are interpreted.

The varieties of cheese that have been associated with outbreaks of staphylococcal food-poisoning are Cheddar and similar types; recently Swiss-type cheese was involved in a Canadian outbreak (Todd et al., 1981). Outbreaks have been infrequent since 1965, perhaps because of a better understanding of the behaviour of *Staph. aureus* in milk before and during cheese-making, which has led to the application of effective preventive measures (ICMSF, 1980b, pp. 510–513).

A viable count of 10,000 *Staph. aureus* per gram of cheese appears to be well within technological achievement, and numbers below that level in ripened cheese in international trade are likely to indicate that numbers had not reached a level associated with detectable toxin (see above comment on reliability). The sampling plan indicated for case 8 would seem to apply (see Table 10, p. 74). However, because of the different rates of die-off of staphylococci in different cheese varieties during ripening, specifying an *m* value in a 3-class plan, which would assure GMP, is not possible. Accordingly, a 2-class plan is recommended for case 8 as shown in Table 24.

The thermonuclease (TNase) test for assessment of staphylococcal growth and possible presence of enterotoxins in cheese is recommended. The utility of this test has been shown by many authors (Cords and Tatini, 1973; Tatini et al., 1975, 1976; Park et al., 1978; van

Schouwenburg–van Foeken et al., 1978; Stadhouders et al., 1980; Todd et al., 1981; Ibrahim and Baldock, 1981). In some instances, however, cheese containing enterotoxin did not contain detectable TNase (Todd et al., 1981), which prompted concern whether cheese suspected of containing enterotoxins, or of having contained large numbers of *Staph. aureus*, should be released for consumption if TNase is not detected. However, Ibrahim and Baldock (1981) and van Schouwenberg–van Foeken et al. (1978) always detected TNase in extracts of cheese containing enterotoxin. Further, Ibrahim and Baldock (1981) detected TNase in cheese whenever the count of *Staph. aureus* reached c. 10^6 per gram.

Although the reasons for the above differences are not clear, the TNase assay used by Ibrahim and Baldock (1981) was more sensitive than that used by others (perhaps as much as 30 times greater than the procedure used by Todd et al., 1981). Differences in sensitivity of enterotoxin assay procedures used and other factors may also contribute. Nevertheless Ibrahim and Baldock (1981) concluded that TNase-positive cheese should be considered unfit for further processing for human consumption unless proved to be free from staphylococcal enterotoxins, and that TNase-negative cheese containing *Staph. aureus* may be processed for human consumption. Similarly, van Schouewenburg–van Foeken et al. (1978) previously concluded that the presence of enterotoxin is unlikely in cheese negative for TNase and that it can be declared safe for consumption. However, cheese positive for TNase may not always contain detectable enterotoxin. Therefore, TNase-positive cheeses must always be subjected to enterotoxin assay before they are declared unsafe for consumption. For further information on staphylococci in cheese and measures for their control, reference to the following should be helpful: IDF, 1973, 1980a; NCI, 1980.

During the latter part of 1971 and early 1972 numerous outbreaks of gastroenteritis occurred in the United States caused by imported soft cheese which contained pathogenic strains of *Escherichia coli* (enteropathogenic, enteroinvasive, or enterotoxigenic) (Marier et al., 1973). A few earlier outbreaks in other countries had been reported (Costin et al., 1964; Gaines et al., 1964). While the coliform problem in cheese is well known, presence of these organisms in many cheese varieties is extremely difficult to prevent completely. With some varieties, if coliforms are present initially, it is virtually impossible to prevent their growth during manufacture or during the ripening period. In several types of cheese *E. coli* can even be considered characteristic. With the exception of some strains of *E. coli* high populations of coliforms are unlikely to present a health hazard. There is ample evidence that if pathogenic strains of *E. coli* (PEC) are present early in the cheese-making process their numbers may increase to hazardous levels (Mourgues et al., 1977; Frank et al., 1977, 1978; Park et al., 1978). However, in view of the scarcity of evidence of recurring outbreaks due to PEC in cheese, and the high cost of routine testing, it is doubtful that establishment of end-product criteria for either coliforms or *E. coli* would be justified. Accordingly, no sampling plan is proposed.

Nevertheless, every effort should be made to adhere to manufacturing practices that will prevent coliform, especially *E. coli*, contamination and subsequent growth (see ICMSF, 1980b, pp. 507, 510, 511). Toward this end the direct plate method (DPM) of Anderson and Baird-Parker (1975) for the enumeration of *E. coli* and as modified by Holbrook et al. (1980) to include a resuscitation step should be useful. With this method results are obtained in 24 hours rather than the 4 days by traditional procedures. Thus, it would be feasible to examine on a routine basis soft and semi-soft cheese for *E. coli* should subsequent developments deem it necessary. Certainly, the application of guidelines for *E. coli* at critical control points of cheese manufacture is advisable, and industry should be encouraged to establish and use them as part of an internal microbiological safety and quality assurance system.

C. SAMPLING PROCEDURES

For general instructions on collecting and handling field samples, see Chapter 10. For specific procedures on sampling milk and milk products consult the IDF Standard 50A proposed by the International Dairy Federation (IDF, 1980c). IDF procedures not included in the general procedures in Chapter 10 are outlined below, but it is recommended that the full IDF text be studied by all persons who collect samples of dairy foods.

(a) *Dried milk*

1 Where feasible, take original unopened packages.

2 When sampling bulk containers in the field or packages in the laboratory, use the precautions described in Chapter 9, Section B (c), when opening the containers. Take the sample from a point near the

centre of the container if possible. First remove the surface layer of powder with a sterile instrument (e.g., a broad-bladed knife) and then draw the sample with a sterile spoon or probe.

3 In the event of dispute concerning the bacteriological condition of the surface powder in a package, take a sample of the product from this area.

4 Preferably use brown glass sample containers, to exclude light.

(b) *Cheese*

1 Use one of the following three techniques, depending on the shape, weight, and type of cheese: (a) sampling by cutting out a sector; (b) sampling by means of a trier; (c) taking a complete cheese as a sample unit. When a choice must be made between (a) and (b), method (b) is often more practicable, especially with hard cheese of large size.

(i) *Sampling by cutting out a sector*. Using a knife with a pointed blade, make two cuts radiating from the centre of the cheese. The size of the sector thus obtained should be such that after removal of any inedible surface layer, the remaining edible portion is more than twice as much as will be required for analysis. Use this method for Edam and Gouda cheese. It can also be used for semi-hard and soft cheese, and in general for all cheese when sampling with a trier is impracticable.

(ii) *Sampling by means of a trier*. The trier may be inserted obliquely towards the centre of the cheese once or several times into one of the flat surfaces at a point not less than 10 to 20 cm from the edge. From the plug or plugs thus obtained cut off not less than 2 cm of the extremity containing the rind and use this piece to close the hole made in the cheese. The remainder of the plug or plugs constitutes the sample unit.

To prevent avoidable loss of product close the plug holes with great care, especially with a large cheese, and if possible seal over with an approved compound. This method is most suitable for hard and semi-hard cheese, such as Emmental and Cheddar.

The trier may be inserted perpendicularly into one face and passed through the centre of the cheese to reach the opposite face. This method is suitable for Provolone, Caciocavallo, and similar varieties.

The trier may be inserted horizontally into the vertical face of the cheese, midway between the two plane faces, towards the centre of the cheese. This method is suitable for cheese such as Tilsit, Cantal, Roquefort, Pecorino, and Romano.

In the case of cheese transported in barrels, boxes, or other bulk containers, or which is formed into large compact blocks, sampling may be performed by inserting a 10 to 15 inch trier obliquely through the contents of the container from the top toward the base. This method is suitable for processed cheese and cheese foods.

(iii) *Sampling by taking an entire or substantial portion of cheese*. Use this method for fresh cheese (for example, cottage, cream, or double cream cheese), for soft cheese of small size, and for wrapped portions of cheese packaged in small containers (e.g., some processed cheese and various soft cheeses).

2 Immediately after sampling, place each sample unit (plug, sector, entire small cheese) in a sterile container of suitable size and shape, and seal. The sample unit may be cut into pieces for insertion into the container, but do not compress or grind it.

3 Send or transport the sample units immediately to the laboratory, and conduct the analysis as soon as possible, preferably on the same day. If either dispatch or analysis must be delayed, place the containers in a refrigerator at a temperature of 5 to 8°C. This is especially important with perishable cheese, such as soft cheese.

Precaution. If bacteriological, chemical, and/or organoleptic analyses are to be made on the same sample unit, always remove the analytical units for the bacteriological tests first.

REFERENCES

APHA (American Public Health Association). 1978. *Standard Methods for the Examination of Dairy Products*, 14th ed. APHA, Washington, DC

Anderson, J.M., and Baird-Parker, A.C. 1975. A rapid and direct plate method for enumerating *Escherichia coli* biotype 1 in foods. *Journal of Applied Bacteriology*, *39*, 111–117

Anderson, P.H.R., and Stone, D.M. 1955. Staphylococcus food poisoning associated with spray dried milk. *Journal of Hygiene*, *53*, 387–397

Armijo, R., Henderson, D.A., Timothee, R., and Robinson, H.B. 1957. Food poisoning outbreaks associated with spray-dried milk. *American Journal of Public Health*, *47*, 1093–1100

Christian, J.H.B. 1983. *Microbiological Criteria for Foods. Summary of Recommendations of FAO/WHO Expert Consultations and Working Groups 1975–1981*. Document VPH/83.54, WHO, Geneva

Codex. 1979. *Codex Alimentarius Commission, Thirteenth Session, 1979: Report of the 15th Session of the Codex Committee on Food Hygiene, Washington, D.C. 18–22 September 1978*. ALINORM 79/13 Appendix V, Code of Hygienic Practice for Foods for Infants and Children. Codex Alimentarius Commission, FAO, Rome

– 1981. *Codex Alimentarius Commission, Fourteenth Session, 1981: Report of*

17th Session of the Codex Committee on Food Hygiene, Washington, D.C. 17–21 November 1980. ALINORM 81/13. Appendix V, Extract from the Report of the FAO/WHO Working Group on the Establishment and Application of Microbiological Criteria for Dried Milk Products and Natural Mineral Waters. Codex Alimentarius Commission, FAO, Rome

– 1983. *Codex Alimentarius Commission, Fifteenth Session, 1983: Report of 18th Session of the Codex Committee on Food Hygiene, Washington, D.C. 22–26 February 1982*. ALINORM 83/13. Appendix II, Draft Code of Hygienic Practice for Dried Milk. Codex Alimentarius Commission, FAO, Rome

Collins, R.H., Trager, M.D., Goldsby, J.B., Boring, J.R., Cohoon, D.B., and Bar, R.N. 1968. Interstate outbreak of *Salmonella newbrunswick* infection traced to powdered milk. *Journal of American Medical Association*, *203*, 838–844

Cords, B.R., and Tatini, S.R. 1973. Applicability of heat stable deoxyribonuclease assay for assessment of staphylococcal growth and the likely presence of enterotoxin in cheese. *Journal of Dairy Science*, *56*, 1512–1519

Costin, I.D., Voiculescu, D., and Gorcea, V. 1964. An outbreak of food poisoning in adults associated with *Escherichia coli* serotype 86:B7:H34. *Pathologia Microbiologia*, *27*, 68–78

Devoyod, J.J., Bret, G., and Auclair, J.E. 1968. The microbial flora of Roquefort cheese. I. Its evolution during cheesemaking and ripening. *Le Lait*, *48* (479/480), 613–629

– 1969. The microbial flora of Roquefort cheese. II. Staphylococci and micrococci. *Le Lait*, *49* (481/482), 1–20

FAO/WHO (Food and Agricultural Organization/World Health Organization). 1970. *Third Report of the Joint FAO/WHO Expert Committee on Milk Hygiene*. WHO Technical Report Series 453. WHO, Geneva

Frank, J.F., Marth, E.H., and Olson, N.F. 1977. Survival of enteropathogenic *Escherichia coli* during manufacturing of Cambembert cheese. *Journal of Food Protection*, *40*, 835–842

– 1978. Behavior of enteropathogenic *Escherichia coli* during manufacture and ripening of Brick cheese. *Journal of Food Protection*, *41*, 111–115

Gaines, S., Achavasmith, Y., Thareesarvat, M., and Dungamani, C. 1964. Types and distribution of enteropathogenic *Escherichia coli* in Bangkok, Thailand. *American Journal of Hygiene*, *80*, 388–394

Holbrook, R., Anderson, J.M., and Baird-Parker, A.C. 1980. Modified direct plate method for counting *Escherichia coli* in foods. *Food Technology in Australia*, *32* (2), 78–83

Ibrahim, G.F., and Baldock, A.K. 1981. Thermostable deoxyribonuclease content and enterotoxigenicity of Cheddar cheese made with sub-normal starter activity. *Journal of Food Protection*, *44*, 655–660

ICMSF. 1978. *Microorganisms in Foods. 1. Their Significance and Methods of Enumeration*, 2nd ed. University of Toronto Press, Toronto

– 1980a. *Microbial Ecology of Foods*. Vol. 1. *Factors Affecting Life and Death of Microorganisms*. Academic Press, New York

– 1980b. *Microbial Ecology of Foods*. Vol. 2. *Food Commodities*. Academic Press, New York

IDF (International Dairy Federation). 1970. *Colony Count of Dried Milk and Whey Powder*. IDF Standard No. 49. IDF, Brussels

– 1971. *Dried Milk and Dried Whey Count of Coliforms*. IDF Standard No. 64. IDF, Brussels

– 1973. *Code of Hygienic Practice for the Manufacture of Cheese*. Document No. 70. IDF, Brussels

– 1980a. *Behaviour of Pathogens in Cheese*. IDF, Brussels

– 1980b. *General Code of Hygienic Practice for the Dairy Industry*. IDF, Brussels

– 1980c. *Milk and Milk Products – Guide to Sampling Techniques*. IDF Standard 50A. IDF, Brussels

ISO (International Organization for Standardization). 1981. *Microbiology – General Guidance on Methods for the Detection of Salmonella*. International Standard 6579. ISO, Geneva (Also appears as BSI Document: BS5763: Part 4: 1982, *Methods for Microbiological Examination of Food and Animal Feeding Stuffs*. Part 4. Detection of *Salmonella*)

Marier, R., Wells, J.G., Swanson, R.C. Callahan, W., and Mehlman, I.J. 1973. An outbreak of enteropathogenic *Escherichia coli* foodborne disease traced to imported French cheese. *Lancet*, *ii*, 1376–1378

Mourgues, R., Vassal, L., Auclair, J., Mocquot, G., and Vandeweghe, J. 1977. Origin et développement des bacteries coliformes dans les fromages à pâté molle. *Le Lait*, *57*, 131–149

NCI (National Cheese Institute). 1980. *Major Policy Recommendations. Microbiological Levels*. NCI, Chicago

Park, C.E., El Derea, H.B., and Rayman, M.K. 1978. Evaluation of the staphylococcal thermonuclease (TNase) assay as a means of screening foods for growth of staphylococci and possible enterotoxin production. *Canadian Journal of Microbiology*, *24*, 1135–1139

Stadhouders, J., Hassing, F., and Galesloot, T.E. 1980. A rapid and simple method for the detection of *Staphylococcus aureus* thermonuclease in cheese. *Netherlands Milk and Dairy Journal*, *34*, 199–204

Tatini, S.R., Cords, B.R., and Gramoli, J. 1976. Screening for staphylococcal enterotoxins in food. *Food Technology*, *30* (4), 64, 66, 70, 72, 73

Tatini, S.R., Jezeski, J.J., Olson, J.C. Jr., and Casman, E.P. 1971a. Factors influencing the production of staphylococcal enterotoxin A in milk. *Journal of Dairy Science*, *54*, 312–320

– 1971b. Production of staphylococcal enterotoxin A in Cheddar and Colby cheese. *Journal of Dairy Science*, *54*, 815–825

Tatini, S.R., Soo, H.M., Cords, B.R., and Bennett, R.W. 1975. Heat stable nuclease for assessment of staphylococcal growth and likely presence of enterotoxins in foods. *Journal of Food Science*, *40*, 352–356

Tatini, S.R., Wesala, W.D. Jezeski, J.J., and Morris, H.A. 1973. Production of staphylococcal enterotoxin in Blue, Brick, Mozzarella and Swiss cheese. *Journal of Dairy Science*, *56*, 429–435

Todd, E., Szabo, R., Robern, H., Gleeson, T., Park, C., and Clark, D.S. 1981. Variation in counts, enterotoxin levels and TNase in Swiss type cheese contaminated with *Staphylococcus aureus*. *Journal of Food Protection*, *44*, 839–848

USPHS, FDA (United States Public Health Service, Food and Drugs Administration). 1978. *Grade A Pasteurized Milk Ordinance, 1978. Recommendations of the US Public Health Service / Food and Drug Administration*. US Government Printing Office, Washington, DC van Schouwenburg–van Foeken, A.W.J., Stadhouders, J., and Jans, J.A. 1978. The thermonuclease test for assessment of the growth of coagulase positive staphylococci in Gouda cheese with a normal acidity development. *Netherlands Milk and Dairy Journal*, *32*, 217–231

Zottola, E.A., and Jezeski, J.J. 1969. Comparison of short-time holding procedures to determine thermal resistance of *Staphylococcus aureus*. *Journal of Dairy Science*, *52*, 1855–1857

Zottola, E.A., Jezeski, J.J., and Al-Dulaimi. 1969. Effect of short-time subpasteurization treatment on the destruction of *Staphylococcus aureus* in milk for cheese manufacture. *Journal of Dairy Science*, *52*, 1707–1714

16

Sampling plans for eggs and egg products

Eggs and egg products are important commodities in international trade. Exports of shell eggs exceed 20 million boxes per annum, and exports of egg products approach 100,000 tonnes. Of the latter, 85–90% is in liquid or frozen form and of this some 60% is whole egg (IEC, 1980). For the purposes of this chapter, eggs and egg products are subdivided into (1) shell eggs and (2) liquid, frozen, and dried eggs. Mention is also made of foods containing eggs.

The introduction to the subject in ICMSF, 1980, p. 521, describes the properties of the whole egg in the shell and the methods of preservation for the three categories of egg products. The whole egg in the shell is self-protective unless abused by excessive exposure to contamination at the laying stage and gross temperature and humidity changes during storage. Immediately after the shell is broken, by whatever means, the liquid egg is exposed to contamination from hands, equipment, and the external surface of the shell itself.

The most important pathogen likely to be present in liquid, frozen, and dried eggs is *Salmonella*. This organism may occasionally be present within the egg at the time of laying, but much more commonly it contaminates the liquid egg from the external surface of the shell during breaking of the shell. There are then the dangers of (a) eating the contaminated raw product, (b) cross-contamination from raw or unpasteurized egg products to foods prepared in the bakery and confectionery trades or in food-service kitchens, (c) direct contamination when an unpasteurized dried-egg product is incorporated into blended foods, and (d) inadequate heating of egg products (e.g., those used for infant feeding) before consumption.

Mandatory pasteurization of liquid egg should eliminate these hazards. However, any salmonellae surviving pasteurization or contaminating the product subsequently may be expected to survive freezing or drying. Thus, salmonellae remain a potential hazard in egg products, because of the opportunities for multiplication that may arise from time/temperature abuse after thawing or rehydration.

Manufactured foods containing egg, such as certain pastas, puddings, and frostings, may also constitute a *Salmonella* hazard. Although prepared with egg of acceptable bacteriological quality, a poorly controlled process may permit substantial multiplication of contaminating salmonellae before the a_w falls below the level limiting their multiplication.

Pasta may be a vehicle for the growth of staphylococci and the formation of enterotoxin. Growth may occur during mixing if wet spots develop and are allowed to persist in the mixer, resulting in detectable levels of enterotoxin in commercial products.

A. SAMPLING PLANS

(a) *Shell eggs*

Sampling plans are not proposed for shell eggs.

(b) *Liquid, frozen, and dried eggs*

The cases for the sampling plans are discussed in ICMSF, 1980, with respect to liquid and frozen eggs (p. 560) and dried eggs (p. 565).

The aerobic plate count (APC) may be regarded as an indicator of the quality of the raw material before processing and also of the possibility of contamination after processing; it may be usefully applied at port-of-entry, manufacturing, and retail levels. In the criteria proposed, the APC is omitted for dried egg albumen because (a) there is always extensive growth of Group D streptococci during desugaring, even when the enzymic method is used; these bacteria are more heat-resistant than organisms of concern such as *Salmonella* and will initiate growth at the pH of egg albumen (c. 9.3); and (b) the pasteurization process that can be applied to liquid albumen must be less severe than that used for liquid whole egg and yolk.

The criteria for APC listed in Table 25 differ slightly from those in ICMSF, 1978, pp. 111–115, 120. A major report (FAO/WHO, 1975), which included a survey of national criteria for egg products, concluded that there was no justification for setting different plans and limits for dried and frozen egg and that the ICMSF (1978) m value of 10^4 was too strict. The APC criteria in Table 25 conform with the recommendation of the above report, which is now part of the Code of Hygienic Practice for Egg Products (Codex, 1978), and with the best estimate based on the data available.

The use of coliform counts is intended for evidence of (a) insufficient heat-treatment or (b) post-pasteurization contamination. This test is more applicable at the point of manufacture than at port of entry, particularly when there is no record of manufacturing procedures. In ICMSF, 1978, coliform tests were included in criteria for dried (p. 111), but not for frozen (p. 120), egg products. They have been included in Table 25 for both products at the level suggested previously for dried egg, as they serve the same purpose in both types of product. This is consistent with Codex, 1978.

The *Salmonella* criteria suggested in ICMSF, 1978, were stricter for frozen egg whites (case 11) (p. 120) than for frozen whole egg (case 10) (p. 120) because the former are often not adequately cooked to destroy *Salmonella*. However, all of the egg products considered in this chapter may have a range of food-ingredient applications that will influence the degree of concern in different ways. Thus the case chosen should be carefully considered in relation to the use intended for each product. For normal routine sampling for *Salmonella* in accordance with intended use, case 10, 11, or 12 would apply (see Table 10, p. 74). The age and condition of the intended consumer should also be considered. The very young, the elderly, and the infirm are more susceptible to infection so more stringent sampling plans should be applied to egg products destined for such customers (see Table 12, p. 78).

The criteria suggested in Table 25 taking the *Salmonella* sampling plan as case 11 are consistent with those of the Codex Alimentarius Commission (Codex, 1978).

Occasionally small numbers of staphylococci may be found in salted yolks, presumably because of the selective nature of the high-salt menstruum. However, no criterion is suggested for staphylococci in this product because there is no evidence of hazard. Few countries include staphylococcal counts in their criteria for egg products (FAO/WHO, 1975).

Physical measurements of efficacy of pasteurization, such as the destruction of α-amylase in whole-egg mélange, are practised with good effect in some countries and in particular in the United Kingdom.

TABLE 25

Sampling plans and recommended microbiological limits for pasteurized liquid, frozen, and dried egg products

Test[a]	Method reference[b]	Case	Plan class	n	c	Limit per gram m	M
APC[c]	118	2	3	5	2	5×10^4	10^6
Coliforms	126–131	5	3	5	2	10^1	10^3
Salmonella,[d] normal routine	163	10	2	5	0	0	–
		11	2	10	0	0	–
		12	2	20	0	0	–
Salmonella,[d] for high-risk population[e]	163	10	2	15	0	0	–
		11	2	30	0	0	–
		12	2	60	0	0	–

NOTE: SAMPLING PLANS AND RECOMMENDED LIMITS FOR FOODS CONTAINING EGG: Sampling plans for indicator organisms need to take account of microbial contributions of non-egg ingredients. Where salmonellae are considered a hazard, the above criteria for salmonellae are appropriate.

a APC, see also ISO 4833 (1978*c*); coliforms, see also ISO 4831, 4832 (1978*a, b*); *Salmonella*, see also ISO 6579 (1981).

b Refers to page numbers in ICMSF, 1978.

c Omit APC for dried egg white.

d The case for *Salmonella* is to be chosen in accordance with whether the expected use of the product will reduce, cause no change in, or increase concern (see Table 10 on p. 74 and ICMSF, 1980, Chapter 19, pp. 521–566).

e For food intended for high-risk populations, the case remains as indicated in Table 10, but *n* is increased to increase stringency.

However, it should be emphasized that, for the amylase test to be a reliable indicator of the adequacy of the process, the pasteurization temperature must reach 64.5°C (for 2.5 minutes) or higher (UK Ministry of Health, 1963).

Some unpasteurized eggs are shipped between countries, and represent a serious hazard as vehicles for the dissemination of salmonellae. The international shipment of unpasteurized egg products should be discouraged and their sale prohibited. No recommended limits are proposed for these products.

(c) *Foods containing egg*

There is a wide range of manufactured foods which contain egg products (ICMSF, 1980, p. 524). It is not possible to generalize in terms of sampling plans for indicator organisms, because of the major contributions that non-egg ingredients may make to these populations. The most important pathogens in such foods are likely to be salmonellae and, where they are considered to be a hazard, the criteria for salmonellae in Table 25 are appropriate, with due regard being given to the case. *Streptococcus pyogenes* and *Staphylococcus aureus* are also of concern, particularly in foods prepared in food-service establishments and homes (Riemann and Bryan, 1979).

B. SAMPLING PROCEDURES

Reference should be made to Chapter 9, and in particular to those sections dealing with the sampling of frozen, liquid, and powdered products. The sample will usually consist of unopened containers, such as hermetically sealed cans or consumer packs (e.g., dehydrated egg products). Select the required number of sample units, which may be individual or composited sample units from individual cans or packs.

(a) *Frozen egg*

Drill diagonally through the frozen material in the can after opening with aseptic precautions. Use a drill or brace and sterile bit, and scrape about 50 g of the drillings into a sterile container with a sterile spoon; then close container and can. Alternatively a modified plastic funnel may be used with the electric drill to collect samples (Adams and Busta, 1970). Thaw the sample unit for not more than 90 minutes at room temperature, or place the required amount of analytical unit directly into an homogenizer with pre-enrichment medium in the case of *Salmonella* testing, or into appropriate diluent for other tests.

(b) *Dried egg*

Open the original container or laboratory container (waxed carton or plastic bag) taking aseptic precautions, and spoon the appropriate amount of analytical unit directly into pre-enrichment medium in the case of *Salmonella* testing, or into appropriate diluent for other tests.

REFERENCES

Adams, D.M., and Busta, F.F. 1970. Simple method for collection of samples from a frozen food. *Applied Microbiology*, *19*, 878 (only)

Codex (Codex Alimentarius Commission). 1978. *Recommended International Code of Hygienic Practice for Egg Products*. Addendum 1-1978 to CAC/RCP 15-1976. FAO, Rome

FAO/WHO (Food and Agricultural Organization / World Health Organization). 1975. *Microbiological Specifications for Foods*. Report of a Joint FAO/WHO Expert Consultation, Geneva, 7–11 April 1975. (Document EC/Microbiol/75 Report 1) FAO, Rome

ICMSF. 1978. *Microorganisms in Foods. 2. Sampling for Microbiological Analysis: Principles and Specific Applications*. University of Toronto Press, Toronto

– 1980. *Microbial Ecology of Foods*. Vol. 2. *Food Commodities*. Academic Press, New York

IEC (International Egg Commission). 1980. Six-Monthly Statistics Bulletin, October 1980, London

ISO (International Organization for Standardization). 1978a. *Microbiology – General Guidance for the Enumeration of Coliforms – Most Probable Number Technique at 30°C*. ISO 4831–1978. ISO, Geneva

– 1978b. *Microbiology – General Guidance for Enumeration of Coliforms – Colony Count Technique at 30°C*. ISO 4832. ISO, Geneva

– 1978c. *Microbiology – General Guidance for Enumeration of Microorganisms – Colony Count Technique at 30°C*. ISO 4833–1978. ISO, Geneva

– 1981. *Microbiology – General Guidance on Methods for the Detection of Salmonella*. International Standard 6579. ISO, Geneva

Riemann, H., and Bryan, F.L. (eds). 1979. *Food Borne Infections and Intoxications*, 2nd ed. Academic Press, New York

UK Ministry of Health. 1963. *The Liquid Egg (Pasteurization) Regulations*. HMSO, London

17

Sampling plans for fish and shellfish

Fish and shellfish are taken in nearly all the salt and fresh waters of the world. The annual catch is estimated at about 71 million tonnes, of which 72% is for direct human consumption (FAO, 1979). The catch is made up of a great diversity of species (800 species are listed by FAO), which range over several phyla from primitive molluscs to marine mammals. A large proportion of the fish catch enters international trade, moving world-wide and involving products from both developing and industrialized countries. This is exemplified by the fact that 162 countries in 1978 exported 32% of their total production of 17.63 million tonnes of seafood (FAO, 1979).

A. PUBLIC HEALTH ASPECTS OF FISH AS HUMAN FOOD

Despite the fact that fishery products are of diverse origin and include many species of animals, the composition of the naturally occurring (autochthonous) microflora is similar in most cases. The microorganisms most commonly encountered are typical of the free-living forms found in water and sediment and rarely include any species of mammalian pathogens (ICMSF, 1980, pp. 575–578). Hence fish caught in waters not polluted by human or animal wastes are free from intrinsic microbiological hazard when handled according to good commercial practice.

Indeed fish and other free-swimming marine animals do not usually carry those organisms generally considered to be typical of the mammalian microflora, including *Escherichia coli*, the 'faecal coliforms,' and enterococci. The presence of human enteric organisms on marine food products is clear evidence of contamination from a terrigenous source.

Fish and shellfish products are a minor source of bacterial food-borne disease in North America, the United Kingdom, and Australia (Todd, 1978), but there is a continuing high relative incidence of bacterial food-borne disease from fish products in Japan and probably in Southeast Asian countries where fish are commonly eaten raw or with little cooking (Sakazaki, 1979). In most cases the aetiologic agent is a vibrio though some botulism due to improperly prepared products also occurs in Japan, Poland, and the USSR and in other North Pacific rim countries including northern Canada and the USA (Hobbs, 1976).

A number of different *Vibrio* species have been implicated in food-borne illness resulting from eating seafood. These include *V. parahaemolyticus, V. vulnificus, V. cholerae, V. fluvialis,* and *V. alginolyticus*. Only *V. alginolyticus* occurs consistently on the living animals, and the actual involvement of this organism in human gastroenteritis is somewhat doubtful. *Vibrio parahaemolyticus* is relatively common in warm-water areas and is a cause of gastroenteritis in coastal regions of India and in Japan and countries of Southeast Asia. Disease due to *Vibrio* species is usually associated with situations where large populations of living *Vibrio* species are present on seafoods, either in the raw state where fish and shellfish are eaten uncooked or in recontaminated cooked products which are stored at temperatures that permit growth. However, infective population levels of vibrios can occur on freshly harvested marine animals, particularly where water quality is poor and temperatures are high. Experiments in the USA in connection with the investigation of cholera from ingestion of cooked crabs suggest that normal cooking practices may not always destroy vibrios (CDC, 1981).

Non-proteolytic strains of *Clostridium botulinum* types E, B, and F occur in small numbers in the intestines of fish and may be transferred to flesh during evisceration or other primary processing. Botulism has been caused by ingestion of botulinal toxin produced in improperly fermented products in Japan, in marine products in arctic North America, in hot smoked fish held at improper temperatures, and in imperfectly canned seafoods (Sakaguchi, 1979).

Raw or processed seafoods are in general excellent substrates for the growth of most common bacterial agents of food-borne disease if held at improper temperatures. It is important therefore to avoid contamination of these foods during preparation and storage and to hold them at chill temperatures.

There is a particular problem in the case of sessile molluscs such as oysters, mussels, and clams which grow in and are harvested from inshore waters most likely to be contaminated by human or animal wastes. These organisms feed by passing large volumes of water over selective mouthparts which strain out small particles from the water. They select and concentrate bacteria from the environment and when held in polluted waters may become dangerous vehicles of both pathogenic bacteria and viruses. Contaminated shellfish may be a source of hepatitis A virus as well as of more generally recognized bacterial enteric pathogens (ICMSF, 1978a, p. 65).

At present the only practical control measure is through bacteriological testing of growing-area waters for coliforms and *E. coli*, supplemented by testing of harvested shellfish for aerobic plate count (APC), coliforms, and *E. coli*; shellfish harvesting is closed when bacterial counts rise unusually (FAO/WHO, 1978). This system has proved to be very effective in countries that use it; and food-borne disease, including viral hepatitis, caused by eating shellfish is almost invariably due to animals taken from unregulated or closed beds. However, poor correlation has been reported between virus incidence in shellfish and indicator bacteria counts in water, and a direct test method for hepatitis A virus would be desirable. In cases where the level of contamination is low, relaying in clean waters or active depuration will effectively clean bivalves of undesirable bacteria although virus and some vibrios may persist. The most effective control is mandatory heat-treatment of bivalve molluscs to destroy bacteria and viruses, but this is often impracticable. In any case shellfish in commerce should be tested for enteric indicator bacteria as an index of bacterial quality although this will not assure safety if the animals are collected from unregulated beds since the test method cannot be expected to give reliable indications of the presence or absence of viruses.

Fish and shellfish poisonings are due to toxins in the flesh of animals either intrinsically present or derived from the food they eat. This is true for puffer fish poisoning (caused by tetradotoxin), ciguatera (caused by ciguatoxin and possibly other toxins), and paralytic shellfish poisoning (PSP) (caused by saxitoxin and other toxins). Scombroid toxin is an exception and is apparently due to products of bacterial growth on tunas, mackerels, and related genera after their death. Histamine is most commonly identified as the toxin. These conditions generally require chemical or biological analysis for toxic substances, and therefore no recommendations for criteria are included. However, individuals and agencies concerned with the safety of seafoods in international commerce should be aware of the potential hazard of the toxicants and should include toxin tests where appropriate.

B. THE BASIS FOR SPECIFIC SAMPLING PROPOSALS

The number of outbreaks of food-borne disease from consumption of seafood during the years since the first volume of the book was published has, apart from chemical fish poisoning, shown little change. A higher incidence of bacterial and viral disease from fish and shellfish continues to occur in India, Japan and Southeast Asian countries than in Europe or North or South America. The microorganisms of principal concern on a world-wide basis appear to be vibrios and *C. botulinum*, with hepatitis A virus and enteric bacterial pathogens important in molluscan shellfish. The more common food-borne pathogens such as *Salmonella, Staphylococcus aureus*, and *C. perfringens*, which have been reported in recent years as causes of fish-borne disease, are not typical environmental contaminants but generally gain access to the food during processing or food-service operations.

Most seafood products are handled and stored under conditions which preclude growth and toxin production by *C. botulinum*. Botulism has been particularly associated with improperly processed canned fish, hot smoked fish, and certain pickled and fermented products (Hobbs, 1976). The increasing use of modified-atmosphere storage to extend the shelf-life of fish may increase the risk of growth of *C. botulinum* (Eklund, 1982). In the case of products at risk, proper processing (e.g., heat-treatments to destroy the organism) and control of storage conditions to prevent germination and outgrowth of *C. botulinum* spores are necessary to ensure safety of the product. The types of *C. botulinum* of greatest concern are the non-proteolytic psychrotrophic types B, E, and F. While *C. botulinum* is recognized to be a significant pathogen associated with fish, no criteria are proposed. The detection of *C. botulinum* on a food does not signal an immediate hazard but only indicates potential risk. Microbial counts or simple determination of the presence of *C. botulinum* spores serve no useful purpose. Even in botulism outbreaks toxin tests, not counts of cells or spores, are necessary to confirm involvement of the suspect food.

For the purposes of applying microbiological criteria, seafood products are grouped as follows:

(a) *Fresh, frozen, and cold smoked fish*, including fresh and frozen (including shipboard frozen) marine and freshwater fish, fish blocks, comminuted fish blocks, scallops, cold smoked fish.
(b) *Precooked breaded fish*, including breaded fillets, portions, fish sticks, and fish cakes.

(c) *Frozen raw crustaceans*, including frozen raw shrimp, prawns, and lobster tails.
(d) *Frozen cooked crustaceans*, including frozen cooked shrimp (peeled or unpeeled), prawns, and lobster tails.
(e) *Cooked crabmeat*: cooked or pasteurized, chilled or frozen, whole crab, or picked meat.
(f) *Molluscs*: fresh and frozen mussels, clams, oysters in shell or shucked.

Note that seafoods not included in these categories are omitted because insufficient microbiological data are available. This grouping differs from that in the previous edition of this book (ICMSF, 1978b) in that some product groups have been combined because the criteria were the same. *Escherichia coli* has been accepted as a better criterion of potentially hazardous contamination than 'faecal coliforms,' and is now used here for fish and shellfish. This change conforms with criteria used for other commodities and makes possible the use of rapid and more objective analytical methods based on membranes (Holbrook et al. 1980; Sharpe et al. 1981).

Vibrio parahaemolyticus, Salmonella, and *Staphylococcus* in (a), (b) and (c) have been made optional criteria to be applied in special circumstances, and their appropriate use is indicated in footnotes to Tables 26 and 27 below.

In general there has been little change in the assignment to cases within the new groupings except that *V. parahaemolyticus* is allocated to case 7, 8, or 9 rather than to case 10, 11, or 12 in accordance with the revision on Table 6 (p. 43). A 3-class plan is now proposed for *V. parahaemolyticus*.

Neither hot smoked fish nor semi-conserved fish products have been included. They have been associated with a number of outbreaks of food-borne illness including botulism, staphylococcal intoxication, and salmonellosis, and in the case of hot-smoked mackerel scombroid poisoning, but there is insufficient evidence that microbiological criteria would contribute significantly to preventing outbreaks attributed to those products.

Dried fish are important to world trade. These are produced in both temperate and tropical regions of the world, often under conditions of poor hygiene, and are frequently shipped with minimum protection against casual contamination. Unfortunately, little has been published on the microbiology of dried fish or their involvement in food-borne illness. Therefore, microbiological criteria are not proposed.

The selection of criteria for seafoods has been made with due attention to variability factors such as goegraphical source and handling, storage, and processing practices (ICMSF, 1980, pp. 567–605; FAO/WHO, 1976, 1978; WHO, 1974) and with consideration of existing national and international criteria (Codex, 1977).

Microbiological criteria for molluscs are dependent on proper control of the growing waters and harvesting process. A list of microbiological criteria used in a number of countries is presented in the Codex Alimentarius Commission's *Recommended International Code of Hygienic Practice for Molluscan Shellfish* (FAO/WHO, 1978), which includes the U.S. standards.*

In selecting criteria in Table 27 we have been guided by actual levels of counts reported in seafood products in international trade – particularly products moving in significant quantity – and the reported occurrence of food-borne disease from various fish products.

A realistic sampling plan should allow the inspecting agency to make the necessary examinations without excessive expenditure of manpower and expense. Concern with expenditure is particularly important with fish and fishery products where there is a large and continuing flow of yearly shipments with a great diversity in the size, weight, and type of package. Often the history of handling, processing, packaging, storage, and transportation is unknown. In addition, the perishability of many fish products requires that tests be completed quickly in order to minimize delay within the usual trade channels. With fish and fishery products, counts at 20 to 25°C are probably the most useful indicators of incipient spoilage (utility) because of the psychrotrophic nature of the spoilage flora. Hence we recommend that the APC be carried out at 25°C. The APC at 35°C is frequently of the order of one-tenth of the count at 25°C (Liston, 1957; Shewan, 1977).

* Maximum APC levels are 100,000 per gram in Denmark and 500,000 per gram in the USA; there are no APC criteria in France, Italy, Netherlands, and the UK.

DENMARK APC (20°C): $n = 10$, $m = 100{,}000$ per gram (2-class); *E. coli* I: $n = 10$, absent in 1/5 ml; *Salmonella*: $n = 10$, absent in 1/5 ml.

FRANCE *E. coli*: $n = 5$–10, absent in 1 ml macerated shellfish or up to 2 *E. coli* per ml for shellfish consumed cooked; *Salmonella*: $n = 5$–10, 25 g tested, no *Salmonella* in 25 g.

ITALY *E. coli*: $n = 10$, <6 *E. coli* per gram (by MPN).

NETHERLANDS *E. coli*: $n = 10$, for hold pending investigational sampling, $m = 4$ per ml, for rejection $m = 15$ per ml.

UK A commonly operated guideline <2 *E. coli* per ml.

USA APC (35°C): <500,000 per gram; faecal coliform: <2.3 per gram.

Although in certain countries Good Commercial Practice (GCP) is commonly attained by plants processing seafoods, no information is available for a significant proportion of producers whose products are in international trade. Although the sampling procedures that are recommended attempt to take cognizance of this fact, the first consideration has been the safety of the consumer. It is, however, important that safety tests (e.g., for *Salmonella*) should not be confused with indications of shelf-life or quality such as the APC. It should be recognized that the bacterial counts on raw fish vary very widely both geographically and because of differences in the duration of shipboard storage, which greatly affects counts on processed products and makes the application of criteria with unreasonably low counts with little scope for variation undesirable. Several items of fish and fishery products in international trade have high *APC*s, but there is no evidence that they have given rise to any corresponding health hazard. Accordingly, the recommended APC limits reflect present accepted practice in the fish industry. The faecal coliforms and *E. coli* are particularly useful as indicators of contamination and mishandling of seafoods since the organisms are absent from fish and crustaceans at the time of capture (except in grossly polluted waters). Moreover, fish and shellfish should be held at temperatures below those which support growth of *E. coli* or faecal coliforms when GCP is followed (see FAO/WHO, 1976, 1978). Thus the presence of detectable numbers of these organisms indicates contamination while relatively large numbers suggest temperature abuse in product handling.

In view of the increasing contamination of inshore marine waters with domestic sewage and hence of the increased probability of contamination of fish and crustaceans taken in such waters, examination for enteric viruses might be as important as for *Salmonella* (Clarke and Chang, 1959; Murphy et al., 1979; Bryan, 1980). However, the lack of appropriate methodology prevents this (Cliver, 1979).

C. SAMPLING PLANS

The sampling plans selected for seafood products are outlined in Table 26. Criteria for APC and *E. coli* counts are recommended in all instances. Staphylococcal criteria are recommended for crabmeat and frozen cooked crustaceans because staphylococcal food-poisoning is a recognized hazard in these products. Additional criteria for *V. parahaemolyticus, Salmonella*, and *Staphylococcus* are recommended for breaded or precooked products, warm-water products, and molluscs.

This procedure is recommended to avoid unnecessary routine application of expensive and time-consuming tests except where the known or suspected origin of the seafood products suggests a hazard may exist.

For economic reasons the number of sample units to be examined, *n*, is limited to five except for *Salmonella* and, in one instance, for *V. parahaemolyticus*. Even five samples would tax the laboratory facilities for routine testing available in many areas which import large quantities of fish. Testing for salmonellae is recommended for some crustaceans and fish harvested from warm-water areas, but should be applied on a selective basis taking into account the origin of and processing applied to the products.

Similarly, while *V. parahaemolyticus* is of widespread occurrence in inshore marine waters, sediments, and marine animals, it is usually present in significant numbers only on seafoods taken from tropical or subtropical waters or from temperate-zone waters in summertime. Routine testing for *V. parahaemolyticus* enumerates both non-pathogenic and enteropathogenic strains, but it is unusual, even in foods implicated in outbreaks of food-borne illness, to find high numbers of the pathogenic serotype.

Except for crabmeat, testing for *Staph. aureus* is also recommended only in special circumstances, because most seafood products carry few staphylococci. Where testing is indicated the recommendations are based on demonstrated hazard for that category of food under the conditions specified. It should be noted that counts on selective media for both staphylococci and *V. parahaemolyticus* are often erratic below, or near, 10^2 per gram (see Table 27, footnote *f*). Reproducibility improves at higher numbers.

An aerobic plate count is recommended for all products because of its usefulness as an indicator of utility and the condition and length of storage of products prior to stabilizing processes such as freezing. Most aquatic animals at the point of harvest have counts in the region of 10^2–10^5 organisms per gram. There are exceptions including some tropical shrimp, molluscs, and freshwater fish. An increase in APC to levels in excess of 10^6 per gram is usually indicative of long storage at chilling temperatures or temperature abuse prior to freezing. Thus APC is indicative of general quality and to a lesser extent of handling and storage procedures.

Escherichia coli testing is also included in all cases, and it is noted that faecal coliform counts may be used instead where this is the preferred method. The usefulness of this test for seafoods was indicated earlier and is based on the fact that these organisms should be absent

Allocation of fish and shellfish products to case

Product	Condition	Test	Case
Fresh and frozen fish[a]	Usually cooked before eating; hazard reduced or low	APC *E. coli*	1 4
Precooked breaded fish products[b]	Usually heated before eating but heating may be insufficient to destroy pathogens; low health hazard	APC *E. coli*	2 5
Frozen raw crustaceans[c]	Usually cooked prior to eating; hazard reduced	APC *E. coli*	1 4
Frozen cooked crustaceans[d]	Little or no cooking before eating; frequently held under conditions after thawing which cause increased moderate hazard	APC *E. coli* *Staph. aureus*	2 5 8
Cooked, chilled, frozen crabmeat[e]	Not cooked prior to consumption; known health hazard; hazard increased	APC *E. coli* *Staph. aureus*	3 6 9
Fresh and frozen bivalve molluscs[f]	Often eaten raw or with light cooking only	APC *E. coli*	3 6

a For fish known to derive from inshore or inland waters of doubtful bacteriological quality, particularly in warm-water areas and where fish are to be eaten raw, it may be desirable to test for *Salmonella* by case 10 and *V. parahaemolyticus* by case 7. Smoked fish may be tested for *Staph. aureus* by case 7.

b *Staph. aureus* may also be tested for by case 8, for cooked fish products likely to be mishandled.

c *Staph. aureus* may be tested in breaded products using case 7. *Salmonella* and *V. parahaemolyticus* tests may be applied to products from warm-water areas and *V. parahaemolyticus* tests in products from temperate regions in summer using case 8, if likely to be eaten raw.

d In view of the storage conditions cases 3, 6, and 9 should apply. However such a load of microbiological testing would be impracticable for most fish-testing laboratories. Also, some survey data indicate that cases 3, 6, and 9 would reject an unacceptably large proportion of the product in international trade. Experience suggests that the criteria recommended above would assist in maintaining hygienic standards. In the case of cooked crustaceans originally harvested from waters and processed in regions of known high environmental hazard, *Salmonella* and *V. parahaemolyticus* may be tested for according to cases 11 and 8 respectively.

e Crabmeat from animals harvested from waters above 15°C may be tested for *V. parahaemolyticus* using case 9.

f I. These criteria are to be used only for molluscs from approved harvesting areas where waters are free from enteric bacteria or virus contamination and there is no significant contamination by toxic metals or toxic and carcinogenic chemicals which may be accumulated by animals. Molluscs from non-approved areas should enter trade only after processing by a treatment to destroy enteric bacteria and viruses and may be tested for dangerous chemicals before distribution.

II. *Salmonella* should be tested for when there is concern for bacteriological safety using case 12, and *V. parahaemolyticus* should be tested for in molluscs from endemic areas and harvested from warm waters using case 9.

TABLE 27

Sampling plans and recommended microbiological limits for seafoods

Product	Test[a]	Method reference[b]	Case	Plan	n	c	Limit per gram or per cm² m	M
Fresh and frozen fish and cold-smoked	APC	115	1	3	5	3	5×10^5	10^7
	E. coli	133[e]	4	3	5	3	11	500
Precooked breaded fish	APC	115	2	3	5	2	5×10^5	10^7
	E. coli	133[e]	5	3	5	2	11	500
Frozen raw crustaceans	APC	115	1	3	5	3	10^6	10^7
	E. coli	133[e]	4	3	5	3	11	500
Frozen cooked crustaceans[c]	APC	115	2	3	5	2	5×10^5	10^7
	E. coli	133[e]	5	3	5	2	11	500
	Staph. aureus	220	8	2[f]	5	0	10^3	–
Cooked, chilled, and frozen crabmeat[d]	APC	115	2[g]	3	5	2	10^5	10^6
	E. coli	133[e]	6	3	5	1	11	500
	Staph. aureus	220	9	2[f]	5	0	10^3	–
Fresh and frozen bivalve molluscs	APC	115	3	2	5	0	5×10^5	–
	E. coli	133[e]	6	2	5	0	16	–

a **For additional tests, conditions of election are given in footnotes to Table 26, p. 189. These tests are described in Table 27 (supplement), p. 191.**

b Refers to page numbers in ICMSF, 1978a.

c Note that although *Staph. aureus* is a moderate hazard, it is infrequently found at high levels and a 2-class plan $n=5$, $c=0$ is recommended to reduce analytical work.

d Refer to note *c* for explanation of *Staph. aureus* case assignment. APC for frozen products only.

e See text.

f A 3-class plan would require *m*-values of about 10^1, which are not detectable with current methodologies.

g In this product APC is not related to hazard.

TABLE 27 (supplement)

Additional tests to be carried out when appropriate (refer to footnotes, Table 26, and see text)

Product	Test	Method reference	Case	Plan	*n*	*c*	Limit per gram or per cm^2	
							m	*M*
Fresh and frozen fish and cold-smoked	*Salmonella*	163	10	2	5	0	0	–
	V. parahaemolyticus	202	7	3	5	2	10^2	10^3
	Staph. aureus	220	7	3	5	2	10^3	10^4
Precooked breaded fish	*Staph. aureus*	220	8	3	5	1	10^3	10^4
Frozen raw crustaceans	*Salmonella*	163	10	2	5	0	0	–
	V. parahaemolyticus	202	8	3	5	1	10^2	10^3
	Staph. aureus	220	7	3	5	2	10^3	10^4
Frozen cooked crustaceans	*Salmonella*	163	11	2	10	0	0	–
	V. parahaemolyticus	202	8	3	5	1	10^2	10^3
Cooked, chilled, and frozen crabmeat	*V. parahaemolyticus*	202	9	3	10	1	10^2	10^3
Fresh and frozen bivalve molluscs	*Salmonella*	163	12	2	20	0	0	–
	V. parahaemolyticus	202	9	3	10	1	10^2	10^3

from the freshly caught raw material. They also provide inferential evidence of temperature abuse in handling of the product.

Attention is drawn to the warning footnote in Table 26 concerning standards for molluscs. The hazard associated with these seafoods when unprocessed and of unknown origin is unpredictable and yet potentially so high that no simple set of criteria can ensure a safe product for the consumer. It is essential that the condition of the growing waters from which the animals are harvested be known before hazard can be assessed. Also be aware of the possibilities of paralytic shellfish poisoning and poisoning from other shellfish toxins in certain areas of the world.

The sampling plans and bacteriological criteria are listed in Table 27. For fresh and frozen fish, *m* for APC has been reduced to 500,000 per gram from 1,000,000 per gram because this value is well within the reach of GCP products in most situations. In a few cases where fish are taken in tropical waters or inshore waters heavily contaminated with bacteria this level may not be attainable. For such situations it is recommended that, if the other bacterial safety indicators are within limits, a high *m* may be accepted, since this criterion relates primarily to utility. The *m* for cooked fish has also been set at 500,000 per gram and it is recognized that this is slightly high since most regulatory agencies set their limit for cooked products at 100,000 per gram. Nevertheless, this figure recognizes the actual situation in international trade, and it has not been shown that products with higher counts are associated with food-borne illness. The *m* for APC for raw crustaceans is set at 1,000,000 per gram, recognizing the high count on much of the shrimp in international trade. However, small pink shrimp from cold northern and southern waters generally have much lower counts, and an *m* of 500,000 per gram might be applicable for them. Cooked crustaceans have been allocated an *m* for APC of 500,000 per gram. Again this is somewhat high for a cooked product but more representative of the world trade situation than 100,000 per gram or less. This and other considerations have led to cases 2, 5, and 8 being proposed (see Table 26, footnote d). The values for molluscs represent a compromise among the standards listed for various countries in the *Recommended International Code of Hygienic Practice for Molluscan Shellfish* (FAO/WHO, 1978). They are probably excessively strict for products to be cooked before eating but, since molluscs are often eaten raw or lightly cooked, the more stringent view has been accepted.

D. SAMPLING PROCEDURES

For general directions on collecting sample units and handling analytical units, see Chapter 9. In routine analysis, collect 5 sample units (see Table 27; 10 are needed in certain cases) from each lot (or consignment) and in each case take at least twice as much food as will be required for the laboratory analysis. Use 0.1% peptone water with salt (ICMSF, 1978a, p. 320) as the diluent for all tests; for fish and fish products, experience has shown that this is the best diluent.

(a) *Iced or chilled raw fish*

1 Whole fish in international trade is normally destined for further processing and is not usually tested bacteriologically. Where whole fish are to enter directly into the retail market, bacteriological tests are normally directed at obtaining estimates of bacterial populations on skin surface for large fish and total bacterial count per unit weight for small fish. Sample units usually consist of single fish in the case of large fish and one or more fish in the case of small fish. Surface samples may consist of swabs taken over a prescribed surface area (e.g., 200 cm^2), or of an aseptically excised area of skin including a minimum of underlying tissues. Shake the swab or skin tissue thoroughly in 10 ml of peptone diluent and proceed with analyses as described in ICMSF, 1978a, p. 115 or 118. Express results in counts per unit area (e.g., per cm^2).

2 For chilled or iced fillets of small fish, take a whole fillet for each sample unit; for large fish such as halibut take a representative portion (e.g., three sections from different parts of the fillet). In the laboratory weigh out an analytical unit of 100 g from each sample unit and proceed with blending, dilution, and analysis as in 1. The initial dilution in blending can range from 1:3 to 1:10.

3 For scallops in ice, collect one sample unit from each of five containers. Weigh out 25 g analytical units from each sample unit and proceed with blending, dilution, and analysis as in 1.

(b) *Frozen fish*

Frozen fin fish products are normally shipped as blocks of whole gutted fish, fillet blocks, comminuted fish blocks, or consumer packages of fillets or other prepared material.

1 Large fish such as salmon, tuna, swordfish, or halibut may be

shipped as frozen whole fish (usually gutted) with or without heads. Sampling and bacteriological testing of such products is difficult and generally is not done. In certain cases imported products of this type are detained until examination of a sample from subsequent processing (e.g., steaking operation) has been shown to be satisfactory. Examination of the processed sample follows the methods given in 2 below.

2 Sample units of wholesale lots of frozen fish may consist of individual fish or entire blocks, cartons, or institutional packages; they may be pieces cut, sawn, or drilled from these larger units (preferably during a primary processing operation in the receiving plant). In the laboratory, thaw sample units for 3 hours at room temperature (20–25°C), overnight in a chilled room (2–5°C), or in sealed plastic pouches under running water, and weigh out 100 g analytical units. Whenever possible the analytical unit should be blended with diluent before becoming completely defrosted. The temperature of the product should not rise above 5°C prior to analysis. Conduct blending, dilution, and analysis as in 1. The initial dilution can range from 1:3 to 1:10.

3 For sampling retail packages of frozen fish, use 100 g analytical units, and treat as in 2 above.

4 Sample and analyse frozen raw or cooked shrimp, prawns, and lobster tails, and frozen raw breaded shrimp (items 3, 4, 5 in Table 27) as described in 2 and 3 above.

(c) *Cooked and cold smoked fish*

Sample and analyse cold smoked fish as described for raw (fresh or frozen) fish above. Sample cooked fish by cutting portions and blending at least 100 g with diluent. Then proceed as for raw fish.

(d) *Molluscs*

For shellstock. Clean off the surface using a stiff brush and running water of potable quality. Open the shell using a sterile instrument to cut the adductor muscle and transfer the shell contents (liquor and tissue) to a sterile container using aseptic technique. Dilute with an equal weight of peptone diluent (ICMSF, 1978a, p. 320) and proceed according to ICMSF, 1978a, p. 115. For the *E. coli* count follow the procedures in ICMSF, 1978a, p. 126 et seq.

For shucked meats. Dilute with an equal weight of peptone diluent and proceed as described above.

REFERENCES

Bryan, F.L. 1980. Epidemiology of foodborne diseases transmitted by fish, shellfish and marine crustaceans in the United States 1970–1978. *Journal of Food Protection*, *43*, 859–868, 873–876

CDC (Centers for Disease Control). 1981. *Foodborne Disease Outbreaks: Annual Summary 1978 revised*, pp. 49–50

Clarke, N.A., and Chang, S.L. 1959. Enteric viruses in water. *Journal of the American Waterworks Association*, *51*, 1299–1317

Cliver, D.O. 1979. Viral infections. In *Foodborne Infections and Intoxications*, 2nd ed., ed. H. Riemann and F.L. Bryan, 299–342. Academic Press, New York

Codex (Codex Alimentarius Commission). 1977. *Draft Code of Practice for Shrimps or Prawns*. (Document CX/FFP7/7). FAO, Rome

Eklund, M. 1982. Significance of *Clostridium botulinum* in fishery products preserved short of sterilization. *Food Technology*, *36* (12), 107–112

FAO (Food and Agriculture Organization). 1979. *Yearbook of Fisheries Statistics, 1978*, *47*, 11

FAO/WHO (Food and Agriculture Organization / World Health Organization). 1976. *Recommended International Code of Practice for Fresh Fish*. (Document CAC/RCP 9-1976). FAO, Rome

– 1978. *Recommended International Code of Hygienic Practice for Molluscan Shellfish*. (Document CAC/RCP 18–1978) FAO/WHO, Rome

Hobbs, G. 1976. *Clostridium botulinum* and its importance in fishery products. *Advances in Food Research*, *22*, 135–185

Holbrook, R., Anderson, J.M., and Baird-Parker, A.C. 1980. Modified direct plate method for counting *Escherichia coli* in foods. *Food Technology in Australia*, *32*, 78–83

ICMSF. 1978a. *Microorganisms in Foods. 1. Their Significance and Methods of Enumeration*, 2nd ed. University of Toronto Press, Toronto

– 1978b. *Microorganisms in Foods. 2. Sampling for Microbiological Analysis: Principles and Specific Applications*. University of Toronto Press, Toronto

– 1980. *Microbial Ecology of Foods*. Vol. 2. *Food Commodities*. Academic Press, New York

Liston, J. 1957. The occurrence and distribution of bacterial types on flatfish. *Journal of General Microbiology*, *16*, 205–216

Murphy, A.M., Grohmann, G.S., Christopher, P.J., Lopez, W.A., Davey, G.R., and Millsom, R.H. 1979. An Australia-wide outbreak of gastroenteritis from oysters caused by Norwalk virus. *Medical Journal of Australia*, *2*, 329–333

Sakaguchi, G. 1979. Botulism. In *Foodborne Infections and Intoxications*, 2nd ed., ed. H. Riemann and F.L. Bryan, 389–442. Academic Press, New York

Sakazaki, J.R. 1979. Vibrio infections. In *Foodborne Infections and Intoxications*, 2nd ed., ed. H. Riemann and F.L. Bryan, 173–209. Academic Press, New York

Sharpe, A.N., Peterkin, P.I., and Rayman, M.K. 1981. Detection of *Escherichia coli* in foods: Indole staining methods for cellulosic and polysulfone membrane filters. *Applied and Environmental Microbiology*, *41*, 1310–1315

Shewan, J.M. 1977. The bacteriology of fresh and spoiling fish and the biochemical changes induced by bacterial action. In *Proceedings of the Conference on the Handling, Processing and Marketing of Tropical Fish*, 51–66. Tropical Products Institute, London

Todd, E.C.D. 1978. Foodborne disease in six countries – A comparison. *Journal of Food Protection, 41*, 559–565

WHO (World Health Organization). 1974. *Fish and Shellfish Hygiene*. WHO Technical Report Series No. 550. WHO, Geneva

18

Sampling plans for vegetables, fruits, and nuts

Vegetables and fruits are important in international commerce mainly as fresh, frozen, dried, or canned products. Ground and tree nuts are also shipped internationally. The range of products is enormous, and with the continued introduction of new varieties it is impossible to discuss all hazards presented by them and all requirements of Good Manufacturing Practice (GMP). Therefore only the broad classes of these commodities are considered, together with yeasts, and single-cell proteins. For further information on these commodities and a detailed discussion of their microbiology see ICMSF, 1980, pp. 606–642.

A. RAW VEGETABLES AND FRUITS

Many vegetables and fruits are not cooked before eating. There is an increasing market for fresh vegetables which are chopped, packaged, then chilled or frozen, and intended to be consumed uncooked in the form of salads. Such preparations are subject to more contamination than are unchopped raw vegetables.

Human and animal enteric pathogens (except soil-borne spore-formers such as *Clostridium perfringens* and *Bacillus cereus*) are usually absent from fresh vegetables at harvest, unless they have been fertilized with human or animal wastes or irrigated with water containing such wastes. A few outbreaks of human disease as a result of eating raw vegetables contaminated by waste water have been reported (Bryan, 1977). Recently an outbreak of listeriosis was caused by contaminated cole-slaw (cabbage) (HWC, 1981). Unclean containers may cause additional contamination as demonstrated by the detection of salmonellae on produce packed in wooden crates formerly used to ship iced poultry (NCDC, 1968). Further details on the pathogens of concern and the products involved are given by ICMSF, 1980.

Raw vegetables intended to be cooked prior to eating are less hazardous, because the pathogens contaminating them, other than bacterial spores, are mostly heat-sensitive. Spores are of little concern as a health hazard in the raw product but may become a problem in the cooked product if it is subjected to time-temperature abuse. Nevertheless, raw vegetables and fruits may introduce pathogens into processing plants and kitchen environments, and contaminate other foods.

Good agricultural practices in growing crops, combined with acceptable hygienic methods during harvesting, packing, and transporting of vegetables and fruits, are far more important than microbiological testing. Therefore, adequate knowledge of fertilization, irrigation, and harvesting and washing practices in the production area is essential.

Vegetables likely to be eaten raw should not be obtained from areas where these practices are known, or suspected, to be unhygienic. While the use of potable water to wash and freshen the harvested vegetables is useful both in commercial practice and in the home, the effects of a water rinse on the level of contamination should not be overestimated. Even thorough commercial washing can remove only c. 90% of the surface flora, which means that 10^5 organisms per gram or cm^2 will often remain on the surface of produce after washing. Residual water may support extensive microbial growth during further storage.

Several appropriate codes of hygienic practice and good manufacturing procedure are available (e.g., Codex, 1983; FDA, 1969).

Routine microbiological examination of raw vegetables and fruits is unlikely to reduce hazards to any great extent and is not recommended. However, tightened inspection may be appropriate on lots suspected of being implicated in food-borne illness, or of having been produced under hygienically unsatisfactory conditions (see Chapter 5 Section O: Tightened Inspection).

B. FROZEN VEGETABLES AND FRUITS

Prepacked frozen vegetables and fruits continue to be popular convenience foods and are in international trade. These products have a commendable microbiological safety record. Blanching, a commonly used procedure, intended primarily to inactivate degradative enzymes, usually kills any vegetative form of pathogens contaminating surfaces of vegetables and fruits. Therefore, the bacteriological condition of such products is primarily a reflection of the hygienic practices employed after blanching. In many instances, however, recontamination of the blanched products occurs.

Although sampling plans for aerobic plate counts (APC) and *Escherichia coli* were recommended in the first edition of this book, the Commission no longer supports the need for the APC recommendation. Without information on the history of a product and its processing conditions, it is impossible to interpret the results of an APC performed at a port of entry. If this information is available (e.g., to industry) appropriate limits can be established and case 1 or 2 would be applicable depending on whether or not the product is to be cooked.

Escherichia coli is usually an indicator of faecal contamination so a criterion (case 5) is proposed for both blanched and unblanched products having a pH value greater than 4.5 (Table 28). Although blanching may diminish the level of contamination, recontamination can occur during subsequent handling. Products having a pH value of 4.5 or less are unlikely to cause problems because the low pH accelerates die-off during frozen storage and inhibits growth in thawed products. A sampling plan for *Salmonella* is not recommended because there is neither epidemiological nor laboratory evidence that *Salmonella* occurs in these frozen products.

C. CANNED VEGETABLES AND FRUITS

Both acid and low-acid canned vegetables and fruits are shelf-stable products whose safety is ensured by a heat process, providing that all canning procedures are in accordance with GMP. No sampling plans are proposed for these products (see Chapter 26: Shelf-Stable Canned Foods).

D. DRIED VEGETABLES AND FRUITS

Information on the microbiology of dried vegetables and fruits is scarce (ICMSF, 1980, pp. 606–642). The flora depends largely on the conditions of growing and harvesting of the particular commodity (see Section A of this chapter) and the processing prior to drying (washing), or other cleaning and blanching. Several products, mainly fruits, such as apricots and prunes, are treated with sulphur dioxide. Although such gassing, spraying, or dipping is primarily intended to stabilize colour, it also considerably reduces the microbial load. The method of drying affects the microflora of products but cannot be relied on to eliminate all pathogenic organisms. Drying of vegetables and fruits is mostly

TABLE 28

Sampling plans and recommended microbiological limits for vegetables, fruits, nuts, and yeast

Product	Test	Method reference[a]	Case	Plan class	n	c	Limit per gram m	Limit per gram M
Frozen vegetables and fruits (pH > 4.5)[b]	*E. coli*	126/131	5	3	5	2	10^{2}[c]	10^{3}
Dried vegetables	*E. coli*	126/131	5	3	5	2	10^{2}[c]	10^{3}
Coconut (desiccated)	*Salmonella*	160[d]						
Growth not anticipated			11	2	10	0	0	–
Growth anticipated			12	2	20	0	0	–
Yeast	*Salmonella*	160[d]	12	2	20	0	0	–

a Refers to page number in ICMSF, 1978, where methods are described. Use analytical unit sizes recommended in the methods.
b pH measured at the time of sampling. Commodities with pH 4.5 or less are not likely to represent a major hazard and criteria are not recommended.
c In the absence of systematic data *m* values are estimates.
d See also ISO 6579 (1981).

combined with heating, but the internal temperature of the product during drying rarely exceeds 35 to 40°C because of evaporation of water. Microbial growth can occur during drying in those internal tissues which still contain sufficient moisture. In the final dried product the water activity is usually below 0.70, and only some moulds and osmotolerant yeasts can multiply.

For the reasons explained in Section B, the Commission recommends that a criterion (case 5) for *E. coli* be applied to dried vegetables and fruits, primarily to monitor GMP (Table 28).

To the best of our knowledge dried vegetables and dried fruits have not caused salmonellosis. Therefore a criterion for *Salmonella* is not recommended. If routine monitoring detects *E. coli*, there may be reason to monitor for *Salmonella*.

E. YEASTS AND SINGLE-CELL PROTEINS

Dried yeast and yeast extract have caused outbreaks of salmonellosis. These products are used as ingredients for dietary foods which are often eaten without prior heating. Therefore limits are recommended for *Salmonella*; case 12 applies (Table 28).

Single-cell proteins – that is, proteins derived from microorganisms – are potentially important replacements of traditional proteins in manufactured foods. However, because their usage is limited, criteria are not recommended. If international trade becomes significant the same criteria as for dried yeasts would apply.

F. NUTS

(a) *Ground nuts*

Peanuts are the only important commodity in this category. The only hazard is the presence of mycotoxins. This is controlled by sorting out damaged nuts by visual inspection or electronic devices. Lots are checked for aflatoxins by chemical procedures or ultraviolet light. No microbiological criteria are suggested. Peanut butter is dealt with in Chapter 22 (Fats and Oils).

(b) *Tree nuts*

Though many varieties of tree nuts move in international trade, only coconut gives cause for concern. The hazard is contamination with

Salmonella. Case 11 is appropriate for products where multiplication of *Salmonella* does not occur; in products where growth may occur case 12 is recommended (Table 28). For other tree nuts, testing for *E. coli* has proved an effective monitor of GMP.

G. SAMPLING PROCEDURES

Appropriate methods for sampling frozen or dried foods are given in Chapter 9 (Collecting and Handling Sample and Analytical Units). For sampling most raw fruits and vegetables a rinse technique should be used as the organisms of concern will be present mainly on the surface. Make a 1/10 dilution (w/v) of the commodity in 0.1% peptone water containing 1% tergitol; shake for 1 minute. Care should be taken not to break the skin of fruits.

If acid fruits are blended, the acids released may destroy the organisms being sought. Counts should be expressed on a per gram basis recognizing that data obtained using the rinse technique for different-sized fruits (e.g., watermelon and red currant) are not comparable because of the large difference in the surface/volume ratios.

REFERENCES

Bryan, F.L. 1977. Diseases transmitted by foods contaminated by wastewater. *Journal of Food Protection, 40*, 45–56

Codex (Codex Alimentarius Commission). 1983. *Recommended International Code of Practice: General Principles of Food Hygiene* (CAC/vol. A – Ed. 1), First Revision (1979). FAO/WHO, Rome

FDA (United States Food and Drug Administration). 1969. Human foods; current good manufacturing practice. *Federal Register, 34*, 6977 (CRF, pt. 128)

HWC (Health and Welfare Canada). 1981. Listeriosis – Atlantic Provinces. *Canada Diseases Weekly Report* 7 (52), 257–259

ICMSF. 1978. *Microorganisms in Foods. 1. Their Significance and Methods of Enumeration*, 2nd ed. University of Toronto Press, Toronto

– 1980. *Microbial Ecology of Foods*. Vol. 2. *Food Commodities*. Academic Press, New York

ISO (International Organization for Standardization). 1981. *Microbiology – General Guidance on Methods for the Detection of Salmonella*. International Standard 6579. ISO, Geneva

NCDC (United States National Communicable Disease Center). 1968. *Salmonella Surveillance*. Report No. 72, National Communicable Disease Center, Atlanta, Georgia

19

Sampling plans for soft drinks, fruit juices, concentrates, and fruit preserves

The microbiological safety and stability of this diverse group of products depend on their formulation (including the use of chemical preservatives, carbonation, low pH values, and pasteurization). For details of preservation requirements see ICMSF, 1980, pp. 643–668. Control should be by application of the HACCP system to raw materials, formulations, processing conditions, and packaging (see Chapter 7: Control at Source).

Examples of such controls would be: (a) measurement of temperature during heat-treatment of pasteurized products; (b) measurement of carbonation levels by routine monitoring of CO_2 pressures in carbonated products; (c) measurement of pH, a_w, product viscosity, and preservative levels in products preserved by the use of chemical preservatives in combination with other formulation parameters; (d) maintenance of good hygienic practices to prevent the build-up of spoilage organisms on the plant equipment and in the production areas.

Most microbiological problems arise because of poor quality of raw materials (e.g., fruit concentrates, sugars, and syrups) and poor process hygiene, which lead to overcoming of the preservation system applied during manufacture by the spoilage organisms. Thus there is a need for routine sampling of raw materials, preventive maintenance of equipment, and monitoring of those parts of the process where microbiological contamination can occur (e.g., during filling and in holding tanks). For the routine monitoring of raw materials simple microbiological tests such as the use of the direct microscopic count, a yeast and mould count, or a standard plate count (using a medium capable of supporting lactic acid bacteria) are most generally applicable. For certain materials more specialized tests, such as the determination of diacetyl and acetylmethylcarbinol in citrus and apple juices and their concentrates, may give a better index of microbiological quality (Murdock, 1968).

It is impossible to recommend specifications for specific types of microorganisms in raw materials as these will very much depend on how they are processed and how they are to be used in the final product. For instance, for a pasteurized juice it is important to control levels of heat-resistant moulds in raw materials whereas in a product preserved by the use of chemical preservatives it is more important to know the level of preservative-resistant microbes. Thus, *Gluconobacter* and related bacteria such as *Acetobacter* are important spoilage organisms in products that contain benzoate and/or sorbate and are distributed in plastic containers. If the differentiation of these organisms is important, the method described by Cirigliano (1982) may be used. A case 1 sampling plan would be appropriate for the examination of materials.

Although routine microbiological testing of end-products is not useful for the direct control of manufacturing operations it may provide useful trend data on the performance of a particular operation over a period of time. For this purpose, once a production line has been established to perform to within commercially acceptable spoilage levels, the numbers of containers that should be examined from a production plant on a daily basis should be between 0.01 and 0.1 per cent of the production units taken from production lines.

Unless prior knowledge indicates that sampling of containers should be concentrated on particular parts of a production period (e.g., during start-up, change in product type, or adjustment of packaging equipment), containers should be sampled at random.

Samples should be incubated appropriately and examined for evidence of microbial spoilage. In most products simple visual inspection for haze, sediment, gas, or mould clumps will suffice. If microbiological examinations are made, the methods used should be the most suitable for the likely spoilage organisms. Most spoilage problems result from the growth of yeasts (principally *Saccharomyces*, but often *Brettanomyces* in carbonated beverages) and/or acid-tolerant bacteria such as *Acetobacter, Gluconobacter, Lactobacillus*, and *Leuconostoc*. The complete identification of spoilage organisms is often useful in identifying the cause of a particular microbiological problem.

REFERENCES

Cirigliano, M.C. 1982. A selective medium for the isolation and differentiation of *Gluconobacter* and *Acetobacter*. *Journal of Food Science, 47*, 1038–1039

ICMSF. 1980. *Microbial Ecology of Foods*. Vol 2. *Food Commodities*. Academic Press, New York

Murdock, D.I. 1968. Diacetyl test as a quality control tool in processing frozen concentrated orange juice. *Food Technology, 22*, 90–94

20

Sampling plans for cereals and cereal products

Cereal grains and their products, such as flours, grits, and meal, are important commodities in international trade and form the staple foods of many communities. There are many cereal-based products and a wide variety of confectionery and bakery products with a cereal as a main component. Some of these products are potential microbial food-borne disease risks by virtue of their cereal components but many more may be risks because of the various animal and other vegetable components they contain. Those products for which the use of microbiological criteria would appear to be of value are listed in Table 29. The microbiology of cereals and their products is discussed in detail in ICMSF, 1980, pp. 669–730.

A. GRAINS AND FLOURS

Cereals include wheat, maize, barley, oats, rye, durum, rice, millet, and sorghum. Protein-rich oil seeds, such as soya beans, and the flours and protein isolates and concentrates derived from them are included in this chapter as their storage, handling, and use pose risks that are somewhat similar to those of cereal grains.

Grains, provided they are harvested when in good condition and rapidly dried to a water activity level preventing microbial growth, and then stored under conditions such that the excessive ingress or movement of water is avoided, have virtually no microbiological risks. However, in practice, these conditions are not always met and mould growth may result. The principal organisms of concern are moulds which produce mycotoxins. Many of the field and storage fungi found on grains and in flours derived from them are capable of producing mycotoxins in incorrectly dried or stored cereal products (see ICMSF, 1980, p. 685).

Their control should be principally at source, that is, by controlling conditions during and after harvesting. It is not recommended that these products be routinely monitored for mycotoxins because routine analytical methods for most mycotoxins are not available. However, where the presence of a mycotoxin of concern (e.g., aflatoxin) has been identified in a particular production area, extensive monitoring of the product may be indicated. For methods of analysis of products for mycotoxins, see Horwitz, 1980.

Levels of mould propagules found in good-quality grains and flours vary between approximately 10^2 and 10^5 per gram depending on the type of grain and the growing, harvesting, drying, and storage conditions (including time of storage). Grains and flours may be variously treated, and this circumstance will affect the levels of moulds found on examination. Thus grains may be washed in chlorinated waters and flours may be heat-treated to change functional properties or treated with a bleaching agent such as chlorine dioxide. All these treatments will reduce mould levels, although none will have a significant effect on mycotoxins, if present. Patent flours, made from part of the endosperm of the grain, will generally contain only low levels of moulds. Whole-meal flours will often contain much higher levels as these include the surface layers of the grain, which are those parts principally contaminated with moulds. Thus when deciding on an acceptable limit for moulds it is necessary to know the conditions to which the flour or grain has been subjected prior to sampling. Without such information, results of analyses cannot be properly interpreted. Providing the processors have this knowledge, mould counts of grains in flours may provide useful information concerning the suitability of these products in various formulations. For this purpose a 3-class sampling plan (case 5) with an M value of 10^5 mould propagules per gram would be applicable. In place of a single criterion, a range of m values is proposed because of the wide range of grains, flours, and related products involved for which neither data are available nor a single Good Commercial Practice (m value) would be applicable.

Salmonellae are recognized contaminants of oil seeds and flours, and isolates and concentrates derived from them (e.g., soya beans, cotton seed, and sunflower seeds). Although these commodities are invariably cooked, or heat-processed, before eating, they may be a source of contamination, and levels of salmonellae should thus be controlled. Because the major problem is with soya products a 2-class sampling plan (case 10) for *Salmonella* is recommended for them in addition to the 3-class plan for moulds (Table 29).

Bacillus cereus is common. It is a soil organism and an unavoidable contaminant of grains and flours. While levels of *B. cereus* can be controlled by correct storage and may be reduced somewhat by cleaning grains, it cannot be eliminated. Two cereal products, fried rice and custard powder, have been frequently associated with food-poisoning, although on occasion other cereal-based products have been incriminated (Gilbert, 1979). *Bacillus cereus* food-poisoning associated with fried rice is the consequence of holding cooked rice for several hours at warm (room) temperatures, or in large containers in refrigerators, prior to frying. The problem with custard also results from temperature abuse of the prepared custard prior to consumption. Such incidents are best prevented by education and training of food-handlers and informative labelling of products (e.g., 'refrigerate after preparation for consumption'). They will not be controlled by monitoring products for *B. cereus* contamination. Thus criteria for *B. cereus* are not recommended except for frozen entrées containing rice or cornflour as a main ingredient, where there is a risk of abuse during manufacture. For this purpose a 3-class sampling plan (case 8) is applicable but because of very limited data on a narrow range of products only a tentative *m* can be proposed. The *B. cereus* selective agar of Holbrook and Anderson (1980) is recommended.

Flours and particularly starches are frequently contaminated with thermophilic spore-forming bacteria which can cause spoilage of canned foods stored above c. 37°C. Thus it is recommended that levels of aerobic thermophilic spores in starches and other cereal products to be used in canned products should not exceed 150 spores per 10 g with an average of no more than 125 spores per 10 g based on the method of analysis specified by the National Food Processors Association (Maunder, 1976). This criterion is not applicable to these commodities when intended for other purposes.

Flours used in the manufacture of bread and pastry products may be a source of specific spoilage microorganisms. Of particular concern in breads are the rope-causing bacteria (mucoid variants of *B. subtilis* and *B. lichenformis*). Various microbiological methods have been proposed for monitoring flours for such organisms but none has been found to be useful in practice. Because the sensitivity of bread to rope formation depends on its formulation and expected storage condition it is useful to carry out a practical baking test to determine whether a

TABLE 29

Sampling plans and recommended microbiological limits for cereals and cereal products

Product	Test	Method reference[a]	Case	Plan class	n	c	Limit per gram m	Limit per gram M
Cereals	Moulds	158	5	3	5	2	10^2–10^4[f]	10^5
Soya flours, concentrates, and isolates[b]	Moulds	158	5	3	5	2	10^2–10^4[f]	10^5
	Salmonella[c]	163[d]	10	2	5	0	0	–
Frozen bakery products (ready to eat) with low-acid or high-a_w fillings or toppings	*Staph. aureus*	220	9	3	5	1	10^2[g]	10^4
	Salmonella[c]	163[d]	12	2	20	0	0	–
Frozen bakery products (to be cooked) with low-acid or high-a_w fillings or toppings (e.g., meat pies, pizzas)	*Staph. aureus*	220	8	3	5	1	10^2[g]	10^4
	Salmonella[c]	163[d]	10	2	5	0	0	–
Frozen entrées containing rice or corn flour as a main ingredient	*B. cereus*	274[e]	8	3	5	1	10^3[g]	10^4
Frozen and dried products	*Staph. aureus*	220	8	3	5	1	10^2[g]	10^4
	Salmonella[c]	163[d]	10	2	5	0	0	–

a Refers to page number in ICMSF, 1978.
b Soya flours, concentrates, and isolates should in addition be examined for *Salmonella* (case 10; plan class 2; $n=5$; $c=0$; $m=0$).
c 25 g sample units should be used; these may be pooled for analysis.
d See also ISO 6579 (1981).
e Preferably the method of Holbrook and Anderson (1980) should be used.
f Best estimate based on limited data. Value will depend on type of grain. In case of flours, value will depend on conditions used for their manufacture.
g Best estimate based on limited data; should be checked against products produced under GCP.

particular flour is suitable for bread manufacture; that is, prepare a small batch of bread in the usual way, store the bread under expected conditions of storage, and check by visual inspection whether or not rope develops (see ICMSF, 1980, p. 717).

Refrigerated doughs and other raw pastry products may sour as a result of growth of lactic acid bacteria present in cereal components. Such organisms occur in flours used for making doughs and grow to high numbers on the dough-making equipment, which is difficult to clean. Microbiological methods have been proposed for checking levels of lactic acid bacteria in cereal products (Hesseltine et al., 1969). However, the potential for souring depends on the formulation and storage conditions, and the numbers of lactic acid bacteria that can be tolerated in particular products can be determined only by practical tests. Problems are avoided by strict attention to process hygiene.

B. FROZEN AND DRIED PASTA PRODUCTS

Pastas are raw dough products made from wheat flour, semolina, farina, water, and other ingredients (e.g., egg) to form a stiff dough usually containing about 30% moisture. The dough is extruded into various shapes and for certain types (e.g., ravioli) may be formed around a meat-containing ingredient. Because of ingredients used, conditions of storage prior to drying, temperatures used for drying, and the slow rate of drying which is necessary to prevent damage to certain types of pasta, these may contain salmonellae and *Staphylococcus aureus*. As these products are to be cooked prior to consumption, case 10 is recommended for testing for salmonellae, and case 8 for *Staph. aureus* (Table 29). A limit of $M = 10^4$ per gram is proposed for *Staph. aureus*. However, if *Staph. arueus* is found in pasta or pasta products in excess of 10^3 per gram additional tests such as the thermonuclease test (or ideally tests for enterotoxins) should be used as *Staph. aureus* may be destroyed during processing but any enterotoxins may not be inactivated. For a suitable method see ICMSF, 1978, p. 242).

Pasta may also be used in made-up dishes such as frozen lasagne, and for such products the same limits should apply.

C. FROZEN FILLED AND TOPPED BAKERY PRODUCTS

Most frozen bakery products are not a food-borne disease risk. Those of potential concern are filled or topped with sensitive ingredients such as products of animal origin (for example, meat pies, dairy-cream-filled doughnuts, pies and cakes, pizzas, and flans) or contain coconut. Abuse may occur after these products leave processing plants. For example, the incidents of botulism in California during 1975 and 1976 involving frozen pot pies resulted from temperature abuse to which the frozen products where subjected after thawing, such as the leaving of unheated pies in ovens overnight. Such gross mishandling of a product cannot be taken into account by any criterion. Limits for *Staph. aureus* and salmonellae are considered appropriate for certain types of frozen, filled, and topped bakery products (see Table 29).

The potential for *Staph. aureus* and salmonellae to grow in non-frozen bakery products depends on the formulation of the fillings and toppings. The presence of these organisms depends mainly on whether or not the sensitive ingredients have been pasteurized and hygienically handled prior to use. Of particular concern are those bakery products that contain unpasteurized ingredients, are ready to eat, and are eaten without heating. No criterion is proposed for products containing fillings and toppings that have water activity levels below 0.85, pH values below 4.5, or with a combination of a_w, pH, and preservatives that prevents growth of these organisms. Such products include jams, jellies, certain artifical creams, acid cheeses, spices, and certain fermented meat products.

The hazard also depends on conditions of use. For products likely to be abused after preparation, stricter limits should be applied (i.e., case 9 for *Staph. aureus* and case 12 for salmonellae).

D. SAMPLING PROCEDURES AND METHODS

Appropriate sampling procedures are described in Chapter 9 (Collecting and Handling Sample and Analytical Units), and the relevant microbiological methods are given in Table 29.

REFERENCES

Gilbert, R.J. 1979. *Bacillus cereus* gastroenteritis. In *Food-borne Infections and Intoxications*, 2nd ed., ed. H. Riemann and F.L. Bryan, 495–518. Academic Press, New York

Hesseltine, C.W., Graves, R.R., Rogers, R., and Burmeister, H.R. 1969. Aerobic and facultative microflora of fresh and spoiled refrigerated dough products. *Applied Microbiology, 18*, 848–853

Holbrook, R., and Anderson, J.M. 1980. An improved selective and diagnostic medium for the isolation and enumeration of *Bacillus cereus* in foods. *Canadian Journal of Microbiology, 26*, 753–759

Horwitz, W. 1980. *Official Methods of Analysis of the Association of Official Analytical Chemists*, 13th ed., AOAC, Washington, DC

ICMSF. 1978. *Microorganisms in Foods. 1. Their Significance and Methods of Enumeration*, 2nd ed. University of Toronto Press, Toronto

– 1980. *Microbial Ecology of Foods*. Vol. 2. *Food Commodities*. Academic Press, New York

ISO (International Organization for Standardization). 1981. *Microbiology – General Guidance on Methods for the Detection of Salmonella*. International Standard 6579, ISO, Geneva

Maunder, D.T. 1976. Thermophilic flat sour sporeformers. In *Compendium of Methods for the Microbiological Examination of Foods*, ed. M.L. Speck, 254–262. American Public Health Association, Washington, DC

21

Sampling plans for spices, condiments, and gums

Spices are dried plants or parts of plants which are used primarily for flavouring, seasoning, and imparting aroma or colour to foods and beverages. For the purposes of this discussion herbs, the aromatic leaves and flowers of certain plants, are included as spices. Condiments are spices and blends of spices which may be formulated with flavour accentuators and potentiators to enhance the flavour of foods. 'Spices and condiments' is the official nomenclature adopted by the International Organization for Standardization (ISO, 1972).

Gums include a wide range of plant and microbial polysaccharides, or their derivatives, which hydrate in either cold or hot water to form viscous solutions or dispersions. In addition to tree exudates, actual gums include seaweed extracts (e.g., agar, alginates, and carrageenan), gums from seed (e.g., locust bean, guar), and microbial gum (e.g., xanthan). As hydrocolloids, gums usually exhibit suspending, dispersing, and stabilizing properties, or they may function as emulsifiers, impart gelling properties, or be mucilaginous. Some gums serve as binders, flocculating agents, film formers, foam stabilizers, mould release agents, or lubricants (Peppler and Guarino, 1984).

A. INTERNATIONAL COMMERCE

Spice-bearing plants are chiefly indigenous to the Asian tropics but are grown commercially in many tropical and some temperate climates (e.g., in Hungary). Herbs are grown in more temperate climates of the Mediterranean, Middle East, North Africa, and North America and widely throughout Europe. India, one of the major spice producers and exporters, accounts for more than 50% of the international trade now estimated at 400 million dollars annually. The United States spice consumption, alone, is around 500 million pounds per year. The spice map of the world continues to change as some sources decline, others are revitalized, and new areas of spice production are developed (Peppler and Guarino, 1984).

The major sources of gums include India, Sudan, North Africa, Iran, Asia Minor, Pakistan, and Mediterranean countries (Peppler and Guarino, 1984).

B. COLLECTION OF SPICES AND GUMS

Spices are produced from botanically diverse plants grown in a wide variety of soils and climates. Depending on the spice, various parts of the plant are used, such as the seed (mustard and nutmeg), the berry (pepper), the fruit (celery seed and nutmeg), the leaves (oregano and bay), the root (horseradish), or the bark (cinnamon) (ICMSF, 1980, pp. 731–751).

As with grains, the primary consideration after harvesting is to dry the plant portion to a water activity level sufficiently low to prevent spoilage. Subsequently, further processing may involve, depending on the spice, such treatments as grinding, pulverizing, cleaning, and sorting. Some of these operations may be carried out at or near the point of harvest; others may be done at a remote site after shipment in international commerce. The microflora of spices reflects the diversity of raw materials from which they are derived, together with the varied and climatic conditions under which the plants are grown. These factors are further influenced by conditions of drying and storage, both of which may be extremely primitive (ICMSF, 1980, pp. 731–751).

Processing methods for many gums are proprietary, but patents and other literature provide some insight. Most tree gums are collected by hand and sorted, graded, packed, and shipped from the growing area. The processor further grades, cleans, mills, and blends the gum. As with spices, the water activity of the crude material is reduced to a level to preclude microbial growth. For example, gum ghatti is sun-dried and pulverized. Carrageenan is extracted from red algae with hot water, filtered and concentrated to about 3% gum content, precipitated with ethanol, dried, and milled (Peppler and Guarino, 1984).

C. MICROBIOLOGY OF SPICES, GUMS, AND HERBS

These products commonly carry large numbers of bacteria and moulds, in the case of herbs mainly of soil origin. Conditions of handling after harvest often permit extensive contamination and microbial growth although drying with heat somewhat reduces microbial numbers. The predominating flora is generally composed of aerobic spore-forming bacteria. Non-spore-forming bacteria, indicator organisms, and pathogens may be found.

The numbers of microbial contaminants can be reduced by treatment with ethylene oxide and to a lesser extent with propylene oxide. Irradiation is also effective in reducing microbial populations (ICMSF, 1980, p. 746).

Bacterial spoilage subsequent to harvesting and drying constitutes no problem, but spoilage by fungi may occur during storage and shipping. Of major concern is the role of contaminating microorganisms as spoilage agents in foods where spices and gums are used as ingredients. Spices contaminated with excessive numbers of spore-forming bacteria may be responsible for the spoilage of canned foods and processed meats (ICMSF, 1980, pp. 747–748). Though spore-forming organisms capable of causing gastroenteritis are found in small numbers in spices, they present no public health concern unless the foods into which the spices are incorporated as ingredients are mishandled to the extent that the small number of *Bacillus cereus* and *Clostridium perfringens*, for example, are permitted to multiply extensively. *Salmonella* is found infrequently, but in a wide variety of spices, and on rare occasions has been responsible for outbreaks of human salmonellosis (Laidley et al., 1974; Lystad et al., 1982; Gustavsen and Breen, 1984).

D. SAMPLING PLANS

In the first edition of this book (p.115) sampling plans and microbiological limits were suggested for spices and gums. In retrospect, it is felt that these recommendations were inappropriate. A considerable portion of these commodities in international commerce would not meet the suggested limits; furthermore, failure to meet the limits might or might not have borne a relationship to food quality or safety.

It is now our recommendation that spices and gums be treated as raw agriculture commodities. Accordingly, the ultimate use of such products will be the determinant. A spice or gum which is unsatisfactory for one purpose may be quite adequate for another. For example, black pepper containing large numbers of aerobic spore-forming bacteria may cause spoilage when used as an ingredient in canned foods or cooked sausage, but it could be used as a condiment on the table for seasoning cooked foods. A manufacturer using spices and gums must determine whether the microbiological quality of these products is satisfactory for the use intended (NCA, 1968). If it is not, his alternatives are to seek another source or to render the available product satisfactory. Treatment with ethylene oxide has been practised extensively for many years but there is some concern over its continued use. Gamma-irradiation may be an effective alternative provided flavours are not impaired.

In commercial practice the microbial and colour problems, and to some extent variability, are overcome by using extracts of the spices. For convenience in use these are often diluted on a carrier such as salt or a finely ground cereal (usually rusk, i.e., ground non-sweetened biscuit). Another approach is to use microencapsulated extracts using gelatin or starch as the encapsulant. Use of these forms is increasing at the expense of 'raw spices.'

REFERENCES

Gustavsen, S. and Breen, O. 1984. Investigation of an outbreak of *Salmonella oranienburg* infection in Norway, caused by contaminated black pepper. *American Journal of Epidemiology*, *119*, 806–812

ICMSF. 1980. *Microbial Ecology of Foods*. Vol. 2. *Food Commodities*. Academic Press, New York

ISO (International Organization for Standardization). 1972. *Spices and Condiments*. Nomenclature finalized draft proposal. TC-34/SC-7, ISO, Budapest

Laidley, R., Handzel, S., Severs, D., and Butler, R. 1974. *Salmonella weltevreden* outbreak associated with contaminated pepper. *Epidemiology Bulletin* (Department of National Health and Welfare, Ottawa), *18* (4), 62

Lystad, A., Aasen, S., and Lassen, J. 1982. Food-borne disease surveillance. Outbreak of *Salmonella oranienburg* infection (Norway). *Weekly Epidemiological Record*, *57*, 329–330

NCA (National Canners Association). 1968. *Laboratory Manual for Food Canners and Processors*, vol. 1, 3rd ed. AVI Publishing Co. Westport, Conn.

Peppler, H.J., and Guarino, P.A. 1984. Spices and condiments. In *Compendium of Methods for the Microbiological Examination of Foods*, 2nd ed., ed. M.L. Speck. American Public Health Association, Washington, DC

22

Sampling plans for fats and oils

Four categories of foods were discussed under fats and oils in ICMSF, 1980, pp. 752–777, namely mayonnaise and salad dressings, peanut butter, margarine, and butter. Product criteria would serve no useful purpose for any of these products except nut butters, including peanut butter.

The safety of mayonnaise and salad dressings is related directly to the pH and acetic acid content of the moisture (continuous) phase of the emulsions. The safety and stability of butter and margarine is primarily dependent on heat-treatment of the ingredients and the stability on emulsification to ensure uniformly small water droplets within the fat (continuous) phase. While microbiological criteria may be effectively applied at various stages in the production of these foods, product criteria, for application at ports of entry, would be of extremely limited value. Other spreads of different composition, however, may be less stable microbiologically.

With peanut and other nut butters microbiological stability is related to reduced water activity in the moisture phase of the product. The greatest hazard is the possibility of contamination with salmonellae, and if these organisms are present they will survive for long periods in a finished product although they would not grow in it. Peanut butter is generally eaten without either heating or a treatment that would destroy microorganisms. Therefore, for *Salmonella*, case 11 would apply (see Table 30), unless peanut butter is to be used as an ingredient in a high-moisture food, in which instance case 12 would apply. Thus 2-class plans are suggested with $n = 10$, $c = 0$ for case 11 and $n = 20$, $c = 0$ for case 12.

TABLE 30

Sampling plans and recommended microbiological limits for peanut butter and other nut butters

Product use	Test	Method reference[a]	Case	Plan class	n	c	Limit per gram m	Limit per gram M
Consumed without heating or other treatment to destroy microbes	*Salmonella*	160[b]	11	2	10	0	0	–
Used as ingredient in high-moisture food	*Salmonella*	160[b]	12	2	20	0	0	–

a Refers to page numbers in ICMSF, 1978.
b See also ISO 6579 (1981).

REFERENCES

ICMSF. 1978. *Microorganisms in Foods. 1. Their Significance and Methods of Enumeration*, 2nd ed. University of Toronto Press, Toronto

– 1980. *Microbial Ecology of Foods*. Vol. 2. *Food Commodities*. Academic Press, New York

ISO (International Organization for Standardization). 1981. *Microbiology – General Guidance on Methods for the Detection of Salmonella*. International Standard 6579. ISO, Geneva

23

Sampling plans for sugar, cocoa, chocolate, and confectionery

I. SUGAR AND RELATED SWEETENERS

Sugar (sucrose) is obtained from sugar cane and sugar beets. It is sold for commercial use in both crystalline and liquid form. The liquid form may be sucrose or invert sugar obtained by hydrolysing sucrose. Corn sweeteners (glucose syrups) are obtained by acid and/or enzymatic hydrolysis of corn starch to dextrose, maltose, and higher polymers. They are sold both in solid form and as syrups. An increasingly important sweetener is high-fructose corn syrup (HFS, also known as isoglucose or isosyrup) made by enzymic conversion of glucose syrup to fructose.

The world production of sucrose and other sweeteners mentioned above in 1980 was 113 million tonnes (mt) of which 5 mt was HFS (Earley, 1980). It is predicted that by 1990 about 150 mt of sweeteners will be produced of which 8.2% will be from hydrolysed corn starch, and most of the corn-based sweeteners will be HFS. Sucrose obtained from various fruits and fructose by hydrolysis of inulin contributes only a minor portion of the world supply of sweeteners.

The concentrations of the liquid sweeteners most commonly used in commerce are listed in Table 31 (Marignetti and Mantovani, 1979/80; Minifie, 1980). Sugar syrups with concentrations of 67% w/w or higher, if properly prepared and stored, are usually not subject to spoilage by surface moulds or osmophilic yeasts.

A. REASONS FOR SAMPLING

Dry refined sugar and dried glucose syrups are remarkably stable, have not been vehicles of food-borne pathogens, and for many industrial uses (e.g., jam, confectionery, bakery products) do not contain mi-

TABLE 31

Concentrations of commercial liquid sweeteners

Liquid sweetener (syrup)	Percentage (w/w)
Sucrose	67–68
Invert sugar	76–77
Glucose (26–65 DE)*	75–82
High fructose (42%, 55%, and less frequently up to 90% of sugar is fructose)	76–77

* DE = dextrose equivalent = the percentage of reducing sugars, on a dry basis, calculated as dextrose.

crobes that will spoil products in which they are used. Some liquid sweeteners may become mouldy or ferment unless stored to preclude condensation on the surface and formation of microenvironments permitting growth of surface moulds or osmophilic yeasts. Similarly, contamination with osmophilic yeasts must be avoided to prevent their contaminating sensitive products (Coleman and Bender, 1957; ICMSF, 1980, pp. 778–821). Dry and liquid sugars used in certain canned foods, and for soft drinks, may introduce sufficient numbers of spoilage microorganisms for some to survive the process and spoil the product. For this reason, sugars used in the canning and soft drinks industries are usually subjected to microbiological testing. Both the microbial limits and the analytical methods have been defined (NCA, 1968; 1972; NSDA, 1975; AOAC, 1972).

B. SAMPLING PLANS

It has been suggested (ICMSF, 1980, pp. 802–803) that case 2 be used for dry sugar that is for purposes other than canning or bottling, and case 3 for both dry and liquid sugar to be used for canning or bottling. The basis for this suggestion was the potential for such sweeteners to introduce spoilage organisms into these classes of products. Additionally, liquid sugars (and other liquid sweeteners) may be spoiled by osmophilic yeasts. On further consideration, however, sampling plans and microbiological criteria are not recommended for the following reasons: (a) sugars and other sweeteners have not been involved in microbial food-borne disease; (b) the microbial content of sugars and other sweeteners is mainly the concern of manufacturers using them as ingredients in other products (see NCA, 1972 for limits suitable for purchase specifications).

II. COCOA AND CHOCOLATE

Cocoa beans are the source of material for a wide variety of cocoa and chocolate products. The annual crop is relatively stable at about 1.5 million tonnes, but the price is sometimes subject to enormous changes within a few years (Minifie, 1980). The combined annual per capita consumption of cocoa and chocolate among 19 countries that consume the bulk of the product varies from 0.6 kg in Portugal to 8.9 kg in Switzerland; the median is 4.3 kg (Minifie, 1980).

A. REASONS FOR SAMPLING

Cocoa beans are harvested, fermented, and dried mostly in technically underdeveloped or in developing countries, frequently under primitive jungle conditions. Most of the dried beans have in the past been processed to cocoa, cocoa butter, and chocolate in more technically advanced countries, where most of the product is consumed. However, there is a trend towards processing an increasing percentage of the crop to intermediate products such as chocolate liquor, cocoa press cake, and cocoa butter in the country of origin to increase local economic return. This practice may cause some concern because one step in processing, roasting, is the main germicidal barrier between a raw agricultural product and the consumer. Unless roasting is adequate and handling of roasted beans well controlled, the intermediate products may be contaminated with *Salmonella*.

Salmonellae are not readily destroyed in chocolate liquor or in confectionery made with cocoa or chocolate liquor because of their increased resistance to heat at the low a_w of such products. In cocoa and in chocolate confectionery, salmonellae may persist for years. Chocolate even appears to protect these bacteria against destruction by acid in the stomach. Small numbers (<1 per gram) of salmonellae in chocolate have caused at least one large outbreak of salmonellosis (Craven et al., 1975; D'Aoust et al., 1975) and larger numbers (about 25 per gram) have caused another outbreak (Gill et al., 1983). *Salmonella*-contaminated cocoa has been associated with a similar large outbreak (Gastrin et al., 1972). Even when no illness has been reported from *Salmonella*-contaminated products, the risk has on occasions prompted their recall (Silliker, 1967; Lennington, 1967; Anonymous, 1978).

Moulds are frequently found in cocoa beans, and it is not uncommon to find mycotoxin-producing moulds and occasionally low levels of mycotoxins in cocoa. However, raw cocoa beans do not readily

TABLE 32

Sampling plans and recommended microbiological limits for cocoa, chocolate, and confectionery

Product	Test	Method reference[a]	Case	Plan class	n	c	Limit m
Cocoa	*Salmonella*	163	11	2	10[b]	0	0
Chocolate and other confectionery	*Salmonella*	163	11	2	10[b]	0	0

a Refers to page number in ICMSF, 1978. Methods as described (pp. 163–172) using non-selective enrichment medium made with skim milk and brilliant green (ICMSF, 1978, p. 324).
b The 25 g analytical units may be composited.

support the production of mycotoxins although they do support abundant growth of mycotoxin-producing fungi (Llewellyn et al., 1978) Therefore, there is always a threat of mycotoxins in cocoa products, but toxin has been found only infrequently and only at low levels.

B. SAMPLING PLANS

The ICMSF (1980, p. 818) has suggested that case 2 be used for aerobic plate counts (APC) and mould counts for cocoa, chocolate, and compound coatings containing cocoa, and that case 5 be applied for coliforms and *Escherichia coli*, and case 11 with $n = 10$ and $c = 0$ for *Salmonella*. However, on further consideration, only *Salmonella*, as the main food-borne hazard in cocoa and cocoa products, need be tested for at ports of entry. The total microbial population and the presence of indicator oganisms are generally controlled by suppliers as part of purchase specifications. Based on the inability of these products to support microbial growth, and their non-involvement in food-borne disease other than salmonellosis, there appears little point in testing them for microbes other than *Salmonella* (see Table 32).

III. CONFECTIONERY

Sugar confectionery is eaten in most countries. Minifie (1980) reported that the annual per capita consumption in 17 countries varied from 1.2 kg in Portugal to 6.6 kg in Ireland, with a median of 3.6 kg.

A. REASONS FOR SAMPLING

Confectionery, because of its low a_w, is generally free from microbial hazards and microbial spoilage problems. The main hazards, when they exist, may come from such ingredients as cocoa, coconut, milk powder, and egg albumen, all of which sometimes contain *Salmonella*, and nuts and cocoa, which may contain mycotoxins. Cocoa has been discussed above and the other ingredients in other chapters. The main microbial spoilage problems are (1) explosion of enrobed preparations of fruit and fondant as a result of growth and gas production by osmophilic yeasts introduced with the fruit, (2) explosion of marzipan from gas produced by osmophilic yeasts introduced with nuts or from the environment, and (3) 'Wasserflecken' (globular areas with watery appearance) in marzipan resulting from growth of moulds (Windisch and Neumann, 1965; Windisch et al., 1978). Moulds may also cause spoilage as a result of condensation of moisture in water-impermeable packages (Jarvis, 1982).

B. SAMPLING PLANS

The ICMSF (1980, p. 821) suggested that case 3 was appropriate for spoilage organisms and case 11 for *Salmonella* in confectionery. Spoilage organisms are mainly the concern of the manufacturer. At port-of-entry sampling for *Salmonella* only, case 11 is recommended (see Table 32).

IV. SAMPLING AND TESTING PROCEDURES FOR COCOA, CHOCOLATE LIQUOR, COATINGS CONTAINING COCOA, CHOCOLATE, AND OTHER CONFECTIONERY

For general directions on collecting and handling samples, see Chapter 9 and ISO, 1981.

Among the many factors to be considered when examining food for *Salmonella*, the anti-*Salmonella* activity in some cocoa which is negated by casein is especially pertinent when isolating this organism from cocoa and combinations of cocoa with milk (e.g., milk chocolate) (Park et al., 1976; Zapatka et al., 1977). In the interest of uniformity, we recommend that reconstituted skim milk with brilliant green be used as the non-selective enrichment medium for cocoa-containing products whether or not they already contain milk. This method appears to be most useful for the products of concern, although it differs in some details from several other recommendations (IOCC, 1977; ICMSF, 1978; FDA, 1978; Wilson et al., 1980).

REFERENCES

Anonymous. 1978. FDA seeks injunction against candy firm. *Food Chemical News*, 3 April, 21

AOAC (Association of Official Analytical Chemists). 1972. Detecting and estimating numbers of thermophilic bacterial spores in sugars – Official first action. *Journal of the Association of Official Analytical Chemists*, *55*, 445–446

Coleman, M.C., and Bender, C.R. 1957. Microbiological examination of liquid sugar using molecular membrane filters. *Food Technology, 11* (7), 398–403

Craven, P.C., Baine, W.B., Mackel, D.C., Barker, W.H., Gangarosa, E.J., Goldfield, M., Rosenfeld, H., Altman, R., Lachapelle, G., Davies, J.W., and Swanson, R.C. 1975. International outbreak of *Salmonella eastbourne* infection traced to contaminated chocolate. *Lancet, i*, 788–793

D'Aoust, J.Y., Aris, B.J., Thisdele, P., Durante, A., Brisson, N., Dragon, D., Lachapelle, G., Johnston, M., and Laidley, R. 1975. *Salmonella eastbourne* outbreak associated with chocolate. *Canadian Journal of Food Science and Technology, 8*, 181–184

Earley, T.C. 1980. High fructose corn syrup in the USA, Europe and Japan. In *Proceedings of the International Sweetener and Alcohol Conference*, 62–69. International Business Conferences and Seminars, Financial Times Conference Organization, London

FDA (United States Food and Drug Administration). 1978. *Bacteriological Analytical Manual*. 5th ed. Food and Drug Administration, Bureau of Foods, Division of Microbiology. Association of Official Analytical Chemists, Washington, DC

Gastrin, B., Kampe, A., Nystrom, K., Oden-Johanson, B., Wessel, G., and Zetterberg, B. 1972. An epidemic of *Salmonella durham* caused by contaminated cocoa. *Lakartidningen*, *69* (46), 5335–5338 (original in Swedish)

Gill, O.N., Sockett, P.N., Bartlett, C.L.R., Vaile, M.S.B., Rowe, B., Gilbert, R.J., Dulake, C., Murrell, H.C., and Salmaso, S. 1983. Outbreak of *Salmonella napoli* infection caused by contaminated chocolate bars. *Lancet, i*, No. 8324, 574–577

ICMSF. 1978. *Microorganisms in Foods. 1. Their Significance and Methods of Enumeration*, 2nd ed. University of Toronto Press, Toronto

– 1980. *Microbial Ecology of Foods*. Vol. 2. *Food Commodities*. Academic Press, New York

IOCC (International Office of Cocoa and Chocolate). 1977. *Microbiological Examination of Chocolates and Other Cocoa Products*. Page 101/B/E-1977. The IOCC and the International Sugar Confectionery Manufacturers' Association, Brussels

ISO (International Organization for Standardization). 1981. *Agriculture Food Products – Layout for a Standard Method of Sampling*. Draft International Standard ISO/DIS 7002. ISO, Geneva

Jarvis, B. 1982. The occurrence of mycotoxins in UK foods. *Food Technology in Australia*, *34*, 508–514

Lennington, K.R. 1967. FDA looks at the *Salmonella* problem – Ever-present threat to the industry. *The Manufacturing Confectioner, 48* (July), 31–33

Llewellyn, G.C., Benevides, J., and Eadie, T. 1978. Differential production of aflatoxin on natural and heat-treated cocoa beans. *Journal of Food Protection, 41*, 785–787

Marignetti, N., and Mantovani, G. 1979/80. Liquid sugar. *Sugar Technology Reviews, 7*, 3–47

Minifie, B. 1980. *Chocolate, Cocoa and Confectionery: Science and Technology*, 2nd ed. AVI Publishing Co., Westport, Conn.

NCA (National Canners Association). 1968. *Laboratory Manual for Food Canners and Processors*. Vol. 1. *Microbiology and Processing*. AVI Publishing Co., Westport, Conn.

– 1972. *Bacterial Standards for Sugar*. NCA, Washington, DC

NSDA (National Soft Drink Association). 1975. *Quality Specifications and Test Procedures for 'Bottlers' Granulated and Liquid Sugar*. NSDA, Washington, DC

Park, C., Rayman, K., Stankiewicz, Z., and Hauschild, A.H.W. 1976. Toxic effect of cocoa on *Salmonella* during the pre-enrichment period. *Proceedings of the Canadian Federal Biological Society*, 162

Silliker, J.S. 1967. The *Salmonella* problem at the candy manufacturer's level. *The Manufacturing Confectioner, 47*, (Oct.), 31–33

Wilson, C.R., Andrews, W.H., and Poelma, P.L. 1980. Recovery of *Salmonella* from milk chocolate using a chemically defined medium and five non-defined broths. *Journal of Food Science, 45*, 310–313, 316

Windisch, S., Kowalski, S., and Zander, I. 1978. Demonstration of osmotolerant yeasts in almonds. *CCB Review for Chocolate, Confectionery and Bakery, 3* (2), 28–29

Windisch, S., and Neumann, I. 1965. Über die 'Wasserflecken' des Marzipans und ihre Entstehung. *Zeitschrift für Lebensmittel-Untersuchung und -Forschung 129*, 9–16

Zapatka, F.A., Varney, G.W., and Sinskey, A.J. 1977. Neutralization of the bactericidal effect of cocoa powder on salmonellae by casein. *Journal of Applied Bacteriology, 42*, 21–25

24

Sampling plans for formulated foods

Commercially prepared ready-to-cook or ready-to-eat foods have been widely accepted throughout the world. When such a product consists of major ingredients from more than one commodity category, it is termed a formulated food. Examples of the multitude of such products are meat and vegetable pot pies, seafood and meat salads, ice cream, infant formulae, dried soups, dessert pies, and certain ethnic dishes including certain pasta products, enchilada, and eggrolls. New products are being developed and marketed, and others discontinued. Consequently it is not possible to provide a meaningful list of all formulated foods.

A wide range of processes are used to prepare these foods, which may be offered for sale to the public as perishable, semi-preserved, or shelf-stable products. Relatively minor changes in formulation, especially the post-processing addition of condiments such as grated cheese, sesame seed, ground spices, or chocolate frosting, may alter the microflora of the product to a degree that different microbiological criteria would be needed for apparently similar products.

For the reasons noted above, the ICMSF makes no pretence of covering the entire range of formulated foods. The purpose of this chapter is to provide guidance and examples for those food industries and regulatory agencies that perceive the need to apply microbiological criteria to particular formulated foods. For information about the basic concepts and principles involved, reference should be made to Part I of this book.

Several commodity chapters in Part II of this book contain examples of formulated foods that are traditionally associated with a particular commodity which is a major ingredient of the product (e.g., ice cream with milk products, pasta with cereal products). Some of the formulated foods mentioned in the first edition of this book have been omitted from this revision because there is insufficient information to determine whether microbiological criteria are useful. Accordingly the types of products included in this chapter are dried soup and gravy mixes, meat pies, salads, selected precooked frozen foods and certain dietetic foods, and foods specifically for infants and children. Other products such as salads and pies are described in ICMSF, 1980, pp. 822–833. For most of those foods the quality of the raw materials is of paramount importance for the quality and safety of the end-product. In such cases, setting criteria for the end-products for the purpose of reducing the eventual hazard to the consumer will be less effective than testing the raw ingredients.

For many formulated foods, the number of microorganisms as determined by the aerobic plate count (APC) is not indicative of adherence to Good Manufacturing Practice (GMP). Moreover, the food will be cooked just prior to eating, thus killing most microorganisms, or will be eaten shortly after preparation, thus avoiding any extended opportunity for microbial growth. For example, certain dried soups and ready-to-cook foods, which will be cooked before eating, contain raw ingredients that may have unavoidably large numbers of microorganisms including coliforms and occasionally *Salmonella*. As the process of manufacture does not always reduce the number of microorganisms in the raw material, their presence in retail samples does not indicate a lack of GMP.

A few commercially processed formulated foods, such as meat pies that are eaten cold, have been incriminated in outbreaks of food-borne disease due to lack of GMP in the factory. Most outbreaks have occurred because of post-processing time-temperature abuse, improper storage, or mishandling by the preparer before serving. Commercial food preparation introduces no hazard that is not present in the domestic situation. However, the magnitude of the hazard is much greater because of the greater number of persons exposed to the lot of the commercial product.

In general the need to establish microbiological criteria for most formulated foods has not been demonstrated. Indeed, in the case of dried soups, the Codex Alimentarius Committee on Food Hygiene decided specifically that such criteria are not justified. However, there are some products that require special attention either because they may contain pathogenic organisms that are not killed before eating, or because they are typically eaten by highly susceptible persons such as those debilitated by age or illness. The latter include foods for infants and children who are in the high-risk category, and the costs involved in controlling the microbial populations mentioned below seem justified. Microbiological end-product specifications for these products were established by the Codex Alimentarius Commission (Codex, 1981).

A. FOODS FOR INFANTS AND CHILDREN

Commercially processed foods for infants and children were not considered in detail in ICMSF, 1980, and some comments concerning their manufacture are necessary before considering sampling plans and limits. Canned baby foods, however, will not be discussed here because they must meet the same requirements for safety as other shelf-stable, sterilized products (see Chapter 26, Shelf-Stable Canned Foods, for discussion of these foods). Three other types of foods are dealt with in the Codex Alimentarius Commission's microbiological specifications for foods for infants and children (Codex, 1979). These are dried biscuit-type products, dried instant products, and dried products requiring heating before consumption.

Plain biscuit-type foods are usually thoroughly baked, and no hazard should exist for these because of the heat process and their low a_w. Those with a coating or a filling require more sophisticated manufacturing practices, but the prior heat-treatment of fillings and coatings should be such that the numbers of vegetative forms of microorganisms are sufficiently reduced. The a_w of the product prevents microbiological multiplication during storage and distribution; moreover, since these products are not reconstituted in water or milk before consumption, risks of mishandling of product are minimal.

Dried instant products range from simple milk powders to very complex mixtures. The milk, with or without additional ingredients, is heat-treated (but not sterilized), concentrated, and then spray-dried, or roller-dried when cereals have been added. The heat-treatment is as mild as possible to maintain high nutritive quality. Dried instant foods include certain dry-blended products which are not heated after blending.

Good manufacturing practices should include careful selection of raw materials, hygienic design of processing equipment, adequate control of the critical points during manufacture, and, of course, continuous good housekeeping. Adherence to these good manufacturing practices will normally lead to products containing very few vegetative microorganisms. The APC may, however, be 10^3 to 10^4 per gram, consisting usually of spores that have survived heat-treatments applied during preparation of the product or its ingredients.

Bacillus cereus, a normal contaminant of milk, can be expected to be present in low numbers. Its mere presence does not constitute a hazard, but if the product is sufficiently abused to permit growth to high numbers, a significant hazard could result.

In principle, there should be very little risk that the product contains heat-sensitive pathogenic organisms. Nevertheless, contamination by *Salmonella* occasionally occurs, sometimes from the use of contaminated ingredients added during dry-mixing or from the environment of the production line. The severity of the hazard presented by these organisms depends on subsequent treatment of the product. If adequately heated the hazard from *Salmonella* is removed. If not heated, this hazard remains and will increase if subsequent abuse of the product occurs.

Mishandling – that is, leaving the reconstituted product at room temperature for periods of more than 1 to 2 hours – may present a risk to the young, old, or debilitated consumer. It should also be taken into account that uncontaminated infant formula may become contaminated during preparation. If prepared in advance, such products should be cooled rapidly in order to prevent bacterial growth.

A third category of foods for infants and children is dried foods which require heat-treatment before consumption. The production of such foods by dry-mixing of non-heat-treated raw materials or with less severe heat processes means that most of these products contain higher numbers of microorganisms than do dry instant foods, and even some heat-sensitive bacteria may be present. However, the presence of these microorganisms in these products is accepted with the full knowledge that cooking will reduce their numbers to non-hazardous levels, making the food safe. Instructions for heating such products should appear on the label.

The advisory end-product specifications of Codex Alimentarius are too strict to be met by all the above categories of products in international commerce (Collins-Thompson et al., 1980). These criteria were established to give guidance on what is considered good microbiological quality. However, only at the factory, when the microbiology of raw material and ingredients and the influence of the process are known, will it be possible to judge if divergence from the end-product specifications truly represents some fault in GMP. Products are present on the market, made according to GMP, which contain higher numbers of aerobic bacteria than stated in the Codex criteria (for dried and instant products, APC: $n = 5, c = 2, m = 10^3, M = 10^4$; coliforms: $n = 5, c = 1, m = 3, M = 20$). Such products may have been made with raw materials containing higher spore counts. Moreover, a few coliforms can sometimes be present without indicating a risk to the health of the consumer. Thus, in order to avoid unnecessary rejection of safe foods in retail markets, or at a point of entry where the history of the product is unknown, the criteria for many of these products will have to be more lenient than the end-product specifications of Codex Alimentarius.

B. SAMPLING PLANS FOR FOODS FOR INFANTS AND CHILDREN AND FOR CERTAIN CATEGORIES OF DIETETIC FOODS

The following considerations are applicable to foods for infants and children and certain categories of dietetic foods (e.g., those intended for consumption by debilitated, or otherwise sensitive, consumers).

The microorganisms to be considered are the mesophilic aerobes, coliforms, staphylococci, and salmonellae. The APC may be used to determine the number of bacterial spores in the raw materials, growth of bacteria at or between critical points, the effect of the applied heat-treatment, growth in accidentally rehydrated products during storage and distribution, and, in general, adherence to Good Commerical Practices (GCP). Since all coliforms should be killed during the heat-treatments applied, this group of microorganisms can be used as indicators for post-process contamination and growth; in short, they can be used for the same purposes as the APC to check adherence to GCP. In dry-mix processes, when no heat-treatment is applied, the coliform count is useful to indicate adherence to GCP in the production, storage, and distribution of the raw materials.

Staphylococci may occasionally be present, but usually only in low numbers, as a result of contamination after drying or during dry-blending (see Chapter 15, Milk and Milk Products). In the production of foods for infants and children rigid quality-assurance programs should be be applied to meet the end-product specifications. A massive contamination with staphylococci without an alert during production or an increase in APC or coliforms is highly improbable, while their presence in low numbers does not indicate a direct public health hazard. Such a hazard would arise only after mishandling of the product after reconstitution. In this case the situation is not different from that described above for *B. cereus*.

In the light of these considerations there is no need for routine examination for *Staphylococcus aureus* in these foods. A similar conclusion was reached in Codex, 1979. However, when excessive contamination with *Staph. aureus* is suspected, case 9 ($n = 10, c = 1, m = 10, M = 10^2$) would apply. When test results or other information indicate that a build-up of *Staph. aureus* has occurred, the product should be considered as unsuitable for the use intended. *Salmonella* is considered a hazard because the minimal infective dose might be low for the intended consumer group, and for products requiring reconstitution the hazard will be increased if subsequent mishandling occurs. For these products sampling plans for *Salmonella* are recommended (Table 33).

For plain dried biscuit no microbiological criteria are suggested because there is no hazard which can be limited by setting such criteria. For coated and filled biscuits, tests for coliforms and *Salmonella* would suffice. Since the condition of use of this product would not increase the hazard, case 5 is selected for the coliform criterion. Case 11 is chosen for *Salmonella* because of (a) the possible presence of an ingredient contaminated with *Salmonella* and (b) the conditions of use, which lead to no increased hazard.

For dried and instant products, a criterion for APC is recommended in addition to those for coliforms and *Salmonella*. Directions for use of these products do not include any microbial destructive treatment (e.g., heating) which can be relied on. As these products must be rehydrated, the possibility of subsequent bacterial growth due to mishandling cannot be excluded. Case 6 is selected for the APC and the coliform counts. For *Salmonella* case 12 remains but, because of the high-risk consumer group, *n* is increased to 60 (see Table 12, p. 78).

For dried products that require heating before consumption case 4 is chosen for APC and coliforms and case 10 for *Salmonella*, because cooking will reduce the hazard.

The limits are based on available published and unpublished data, and take into consideration that maintaining the nutritive quality of such products is more important than reducing the numbers of mesophilic bacteria of little public health significance. In the last two categories values of *m* and *M* have been modified from those in Codex, 1981, as a consequence of bacteriological surveys and commercial experience.

TABLE 33

Sampling plans and recommended microbiological limits for foods for infants and children and certain categories of dietetic foods

Product	Test	Method reference[a]	Case	Class plan	n	c	Limit per gram m	Limit per gram M
Coated or filled, dried shelf-stable biscuits	Coliforms	126–131 (Method 3)[b]	5	3	5	2	10	10^2
	Salmonella[e,f]	163–172[c]	11	2	30	0	0	–
Dried and instant products requiring reconstitution	APC[g]	115–118 (Method 1)[d]	6	3	5	1	10^4	10^5
	Coliforms	126–131 (Method 3)[b]	6	3	5	1	10	10^2
	Salmonella[e]	163–172[c]	12	2	60	0	0	–
Dried products requiring heating to boiling before consumption	APC	115–118 (Method 1)[d]	4	3	5	3	10^5	10^6
	Coliforms	126–131 (Method 3)[b]	4	3	5	3	10	10^2
	Salmonella[e]	163–172[c]	10	2	15	0	0	–

a Refers to page number in ICMSF, 1978.
b See also ISO 4831 (1978a).
c See also ISO 6579 (1981).
d See also ISO 4833 (1978b).
e Use 25 g analytical units which may be pooled.
f Applies only to products containing a *Salmonella*-sensitive ingredient.
g Not applicable to fermented products.

C. SAMPLING PROCEDURES

Reference should be made to those sections of Chapter 9 (Collecting and Handling Sample and Analytical Units) dealing with the sampling of dried and powdered products.

In the examination of dried products for *Salmonella*, rapid rehydration by shaking the powder with the pre-enrichment fluid should be avoided (van Schothorst et al., 1979; Andrews et al., 1983).

REFERENCES

Andrews, W.H., Wilson, C.R., and Poelma, P.L. 1983. Improved *Salmonella* species recovery from non-fat dry milk pre-enriched under reduced rehydration. *Journal of Food Science, 48*, 1162–1165

Codex. 1979. *Codex Alimentarius Commission. Thirteenth Session, 1979: Report of the 15th Session of the Codex Committee on Food Hygiene, Washington, D.C. 18–22 September 1978.* ALINORM 79/13. Appendix V, Code of Hygienic Practice for Foods for Infants and Children (up to three years). Codex Alimentarius Commission, FAO, Rome

– 1981. *Codex Alimentarius Commission, Fourteenth Session, 1981: Report of the 17th Session of the Codex Committee on Food Hygiene, Washington, D.C. 17–21 November 1980.* ALINORM 81/13. Appendix VII. Microbiological Specifications for Foods for Infants and Children. Codex Alimentarius Commission, FAO, Rome

Collins-Thompson, D.L., Weiss, K.F., Riedel, G.W., and Charbonneau, S. 1980. Microbiological guidelines and sampling plans for dried infant cereals and powdered infant formula from a Canadian national microbiological survey. *Journal of Food Protection, 43*, 613–616

ICMSF. 1978. *Microorganisms in Foods. 1. Their Significance and Methods of Enumeration*, 2nd ed. University of Toronto Press, Toronto

– 1980. *Microbial Ecology of Foods*. Vol. 2. *Food Commodities*. Academic Press, New York

ISO (International Organization for Standardization). 1978a. *Microbiology – General Guidance for the Enumeration of Coliforms – Most Probable Number Technique at 30°C.* ISO 4831–1978. ISO, Geneva

– 1978b. *Microbiology – General Guidance for Enumeration of Microorganisms – Colony Count Technique at 30°C.* ISO 4833–1978. ISO, Geneva

– 1981. *Microbiology – General Guidance on Methods for the Detection of Salmonella*. International Standard 6579. ISO, Geneva

van Schothorst, M., van Leusden, F.M., de Gier, E., Rijnierse, V.F.M., and Veen, A.J.D. 1979. Influence of reconstitution on isolation of *Salmonella* from dried milk. *Journal of Food Protection, 42*, 936–937

25

Sampling plans for natural mineral waters, other bottled waters, process waters, and ice

A. DEFINITIONS AND ECONOMIC IMPORTANCE OF WATERS IN INTERNATIONAL TRADE

Natural mineral water comes from underground water-bearing strata and is characterized by its mineral salt and trace element constituents. It is subjected to only minimal treatment such as carbonation, aeration, sedimentation, and filtration. A more complete definition is cited by Codex, 1978; a portion of this definition is given below (extracted from the Draft European Regional Standard for Natural Mineral Waters).

2.1 *Definition of Natural Mineral Water*. Natural mineral water is a water clearly distinguishable from ordinary drinking because: (a) it is characterized by its content of certain mineral salts and their relative proportions and the presence of trace elements or of other constituents; (b) it is obtained directly from natural or drilled sources from underground water bearing strata; (c) of the constancy of its composition and the stability of its discharge and its temperature, due account being taken of the cycles of natural fluctuations; (d) it is collected under conditions which guarantee the original bacteriological purity; (e) it is bottled close to the point of emergence of the source with particular hygienic precautions; (f) it is not subjected to any treatment other than those permitted by this standard; (g) it is in conformity with all the provisions laid down in this standard.

2.2 *Supplementary Definitions*

2.2.1 *Naturally carbonated natural mineral water*. A 'naturally carbonated natural mineral water' is a natural mineral water which, after possible treatment in accordance with sub-section 3.1.1 and replacement of gas and after packaging, has the same content of gas from the source as at emergence of the natural mineral water taking into account the usual technical tolerance.

2.2.2 *Non-carbonated natural mineral water.* A 'non-carbonated natural mineral water' is a natural mineral water which, by nature and after possible treatment in accordance with sub-section 3.1.1 and after packaging, does not contain free carbon dioxide in excess of the amount necessary to keep the hydrogen carbonate salts present in the water dissolved.
2.2.3 *Decarbonated natural mineral water and natural mineral water fortified with carbon dioxide from the source.* A 'decarbonated natural mineral water' or a 'natural mineral water fortified with carbon dioxide from the source' is a natural mineral water which, after possible treatment in accordance with sub-section 3.1.1 and after packaging, does not have the same carbon dioxide content as at emergence.
2.2.4 *Carbonated natural mineral water.* A 'carbonated natural mineral water' is a natural water which, after possible treatment in accordance with sub-section 3.1.1 and after packaging, has been made effervescent by the addition of carbon dioxide from another origin.
3.1.1 Treatments permitted include separation from unstable constituents by decantation and/or filtration, if necessary accelerated by previous aeration.
3.1.2 The treatments provided for in sections 2.2.1, 2.2.2, 2.2.3, 2.2.4 and 3.1.1 above may only be carried out on condition that the mineral content of the water is not modified in its essential constituents, which give the water its properties.

Natural mineral water may be either non-carbonated or carbonated. The latter may be either naturally carbonated at its origin or by adding carbon dioxide (CO_2) from its own source or from other sources. Hence, the CO_2 concentration may vary considerably. The pH of carbonated mineral waters is often rather low, between 3 and 4. Natural mineral waters without, or with very little, carbon dioxide are known in Europe as 'flat waters' (*eaux plates*).

Although specific figures are not available, bottled mineral waters are shipped internationally in considerable quantities. Water from similar sources, but not having any particular specified amount of mineral content, and potable* water from other sources are sometimes also bottled. These bottled waters may be sold in large quantities in certain

* Potable means that water must be free from microorganisms capable of causing disease and free from either minerals or organic substances that can produce adverse physiological effects. It must also be aesthetically acceptable and thus be free from apparent turbidity, colour, odour, and objectionable taste and have a reasonable temperature for its intended purpose. Hence, it can be ingested or used in food-processing in any amount without concern for adverse effects on health (Borchardt and Walton, 1971).

countries or regions of a country, but are at present of limited importance in international trade.

Water is used widely in the food industry. It is used to move products, to wash vegetables, fruits, fish, and poultry, to freshen raw vegetables after harvesting and during distribution, and as an ingredient in some foods. Virtually all frozen foods carry a glaze of ice which is often derived from process water, and for certain frozen foods (such as fish and shellfish) a glaze is added as a protective measure. This process water is thus an integral part of some processed foods that are shipped internationally. For example, fish is typically shipped in ice; poultry may be cooled in water and slush ice and transported in ice.

B. REASONS FOR SAMPLING

Even the most protected and best drawn waters are never sterile and possess a certain natural microbial (autochthonous) flora which is rather constant for a particular supply. Subsequent contamination during catching, storing, transporting, or bottling accounts for the introduction of microbial species (allochthonous flora) normally present (ICMSF, 1980b, pp. 834–837).

(a) *Bottled waters*

Contaminated surface or ground (well or spring) waters have frequently been identified as vehicles in outbreaks of water-borne diseases (Holden, 1970; Mossel et al., 1973, 1977; WHO, 1979, 1980). Particular care, therefore, is needed to obtain water from sources that are potable and to protect the sources and water in distribution from contamination. This is particularly important for natural mineral waters because, by definition, they cannot be disinfected before or after bottling and can be subjected only to certain treatments.

Water ingested between meals passes rapidly through the stomach because the pyloric area is relaxed at that time, and thereby avoids the bactericidal effect of intragastric retention (Levine and Nalin, 1976; Mossel and Oei, 1975). Thus, water containing even small numbers of pathogens can cause disease.

Bottled waters have been implicated in outbreaks of water-borne disease, but there is little published information. Bottled, non-carbonated mineral water was a vehicle of *Vibrio cholerae* in a cholera outbreak in Portugal (Blake et al., 1977). Buttiaux (1960) attributed outbreaks of gastroenteritis in Hungary and of typhoid fever among

Polish children to ingestion of bottled water. Persons from countries with a high standard of hygiene who travel to countries where standards are lower often drink bottled water in the hope of avoiding illness. However, they frequently develop travellers' diarrhoea, and sometimes the bottled water comes under suspicion (Gangarosa et al., 1980).

Bottled natural mineral water is sometimes used for domestic preparation of either infant formulas or dietetic foods on the assumption that it is safer than other waters. Consumers of these particular products are often highly susceptible to disease. *Pseudomonas aeruginosa*, for example, may be a hazard for infants (Panero et al., 1970; Mossel, 1977). An epidemic of *P. aeruginosa* infection occurred among infants in a nursery when well water was contaminated by seepage of sewage and by infiltration of stream water (Weber et al., 1977).

Thus, it is essential that the initial bacteriological quality of the water be high and maintained by strict control at its source and by monitoring of the bottling operation. In the absence of information on the efficacy of control at source, importing countries should seek assurance that bottled water is safe by appropriate microbiological monitoring of incoming lots.

(b) *Process waters*

Processing water and ice can be important sources of spoilage organisms and, if from non-potable sources or subsequently contaminated, potential sources of pathogens. They can introduce these organisms to foods during processing, storage, or preparation. If illness occurs as a result of such contamination, the contaminated food is considered as the vehicle, although the initial source of pathogens was process water. For example, raw untreated river water was used for washing fish that were later identified as a vehicle in an outbreak of salmonellosis (Gangarosa et al., 1968). Contaminated cooling water was implicated as the source of *Salmonella typhi* in canned corned beef responsible for several outbreaks of typhoid fever (Howie, 1968). Water used to keep tofu (oriental soya-bean curd) moist in packages was the source of *Yersinia enterocolitica* that was responsible for an outbreak of gastroenteritis (Nolan et al., 1982).

C. SAMPLING PLANS

Unfortunately, no single organism, or group of organisms, serves as an ideal indicator of pollution of water. Tests for the following are sometimes used to evaluate the safety of water: APC (42°C, 37°C, and

20°C), coliforms, faecal coliforms, *Escherichia coli*, Enterobacteriaceae, faecal streptococci, sulphite-reducing clostridia, *E. coli* phages, and specific pathogens. Many of these tests are used to examine bottled water in several countries (Morisetti, 1979). The decision as to which organisms should be sought to evaluate the potability of bottled waters is complicated because unique advantages and disadvantages are associated with each indicator organism or group. These have been reviewed by Cominazzini (1978), Geldreich (1978), Hoadley and Dutka (1976), and Mossel (1976).

(a) *Non-carbonated waters*

Although water-testing laboratories in some countries use the *E. coli* test for routine examination of water for the protection of public health, others, as recommended by WHO (1971) and an FAO/WHO Working Group (1981), use the coliform test. The coliform test, even with its disadvantages, is the method chosen by the Codex working group for examining natural mineral water (Codex, 1978). Microbiological data available to the Commission indicate that bottled natural mineral water can be produced without contamination by coliforms. Examination for coliforms provides a more stringent microbiological rejection criterion than testing for *E. coli*. Routine testing of bottled waters for both coliforms and *E. coli* offers little advantage and increases the cost of analysis.

Cases 4 to 6 apply when indicator organisms are sought. Although certain microorganisms multiply in bottle-stored water, the organisms usually employed as indicators decrease in number with time of storage (Schmidt-Lorenz, 1976). Because time of bottling is often unknown, case 5 seems the most appropriate. This case calls for a sampling plan of $n = 5$ and $c = 2$ (Table 10, p. 74) but such a plan is too lenient for a commodity of special public health concern. These concerns are: (1) the sampling plan employed should give at least the same protection of public health as those required for community water supplies; (2) bottled water is usually stored longer than water in community water supplies, and during the storage period certain indicators of faecal pollution (such as coliforms) may die off more rapidly than some pathogens, while autochthonous or other allochthonus organisms multiply; and (3) home-prepared formulas for infants are sometimes reconstituted with bottled mineral water. These formulas are sometimes subject to temperature abuse between preparation and feeding (e.g., when a child falls asleep during feeding and later wakes and feeds again). A 2-class sampling plan for coliforms with $n = 5$ and $c = 0$ and a limit of $m = 0$ is recommended for examination of bottled non-carbonated waters including natural mineral water (Table 34). Analytical units of 250 ml should be examined for coliforms by the membrane filter method. Decreasing the c value from 2 to 0 and increasing the size of the sample unit to 250 ml make a stringent plan to serve the above-mentioned considerations.

Pseudomonas aeruginosa, a frequent contaminant of the environment, may cause water-borne disease (Hoadley, 1976, Young, 1977, Weber et al., 1977). After adaptation, it can multiply in a mineral water of extremely low nutrient level (Schmidt-Lorenz, 1976). Indicator organisms are of little value in detecting the presence of this opportunistic pathogen. The finding of *P. aeruginosa* in natural mineral water should be cause for concern, particularly if the water is to be consumed by highly susceptible persons such as infants. Therefore, it is recommended that natural mineral water for such a purpose be tested for *P. aeruginosa*. Case 8 applies, and a 2-class plan with $m = 0$ is suggested (Table 34).

Aerobic plate counts (at 35–37°C for 24 hours and at 20–22°C for 72 hours) at source and within 12 hours after bottling may be considered, according to the general principles of FAO/WHO, as guidelines useful in interpreting water quality at its source and its place of bottling. To be acceptable, counts should not exceed those set for natural mineral water by EEC (1980a). APC at 20°C may be useful for quality control within certain food-processing plants.

(b) *Carbonated waters*

Pathogenic organisms (other than spores) should succumb rapidly (usually within hours) after exposure to a pH value of 3.5 or lower (ICMSF 1980a, pp. 92–111). Therefore, it is not necessary to test carbonated natural mineral waters, or other carbonated bottled waters having a pH value of 3.5 or lower, for indicator organisms. To determine whether testing is necessary, a sampling plan that includes testing the pH of five bottles (n) from a lot is recommended. If the pH of any sample unit exceeds 3.5, the sampling plans for testing non-carbonated bottled waters should be followed (see Table 34). If it is not possible to test the pH of the water, the sampling plan for non-carbonated waters should be followed.

TABLE 34

Sampling plans and recommended microbiological limits for bottled water

Product	Test	Method reference	Case	Plan class	n	c	Limit per 250 ml m	Limit per 250 ml M
Non-carbonated natural mineral waters (*eaux plates*) and bottled non-carbonated waters, not classified as mineral waters	Coliforms	MF[a]	5	2	5	0	0	–
	Pseudomonas aeruginosa[b]	See text	8	2	5	0	0	–
Carbonated waters (natural mineral or other bottled waters)	pH	See text		2	5	0	3.5[c]	–

a MF = membrane filter method (see Greenberg et al., 1981). N.B. Resuscitate prior to recovery on selective media (Morisetti, 1979).

b This test is used only if it is anticipated that bottled waters are to be used in infant formulae or for other highly susceptible persons.

c If any sample unit is greater than pH 3.5, proceed with the above sampling plans for non-carbonated waters.

(c) *Process waters and ice*

Guidelines for microbiological criteria of process water and ice must be applied at the processing or manufacturing plant, at their sources, and/or at appropriate distribution points. Process water and ice should be made from potable water, and should meet the microbiological criteria cited for potable water by EEC (1980b), Greenberg et al. (1981), or WHO (1971). Ice glaze is considered as part of the frozen food itself and will be tested during analysis of the food.

D. SAMPLING PROCEDURES

Sample units of bottled water should be collected randomly from the lot as described in Chapter 9 and brought to the laboratory within 1 hour of collection. If this is not practicable, they should be cooled to 10°C or below, iced, and transported to the laboratory in insulated containers. Water should be collected at its source or during processing, distribution, or storage as described by Bryan et al. (1979). When sampling waters at their source, samples should be stored chilled and APCs made within 12 hours. Ice should be brought frozen to the laboratory, either as unopened bags of ice or as ice sampled in bulk with a sterile pick, scoop, or tongs and put into sterile bags or jars.

E. TEST PROCEDURES

The coliform test by the membrane filter technique is recommended because it facilitates sampling large volumes of water. Laboratories not able to use this technique should test for coliforms using the multiple tube dilution (MPN) method. Test procedures for water and ice are not described in ICMSF, 1978, but those given in *Standard Methods for the Examination of Water and Wastewater* (Greenberg et al., 1981) are applicable. Carbonated waters should be degassed before microbiological testing. The pH should be measured by the method specified in Greenberg et al. (1981). There is no standard method for detection of *P. aeruginosa* in water, but methods are given by Greenberg et al. (1981), Mossel et al. (1976), and WHO (1980).

REFERENCES

Blake, P.A., Rosenberg, M.L., Costa, J.B., Ferreira, P.S., Guimaraes, C.L. and Gangarosa, E.J. 1977. Cholera in Portugal, 1974. I. Modes of transmission. *American Journal of Epidemiology, 105*, 337–343.

Borchardt, J.A., and Walton, G. 1971. Water quality. In *Water Quality and Treatment:*

A Handbook of Public Water Supplies, 3rd ed., ed. American Water Works Association. McGraw-Hill, New York

Bryan, F.L., Anderson, H.W., Baker, K.J., Craun, G.F., Duel, W., Lewis, K.H., McKinley, T.W., Robinson, R.A., Swanson, R.C., and Todd, E.C.D. 1979. *Procedures to Investigate Waterborne Illness*. International Association of Milk, Food, and Environmental Sanitarians, Ames, Iowa

Buttiaux, R. 1960. La surveillance bactériologique des eaux minérales en bouteilles et en boîtes. *Annales de l'Institut Pasteur de Lille, 11*, 23–28

Codex (Codex Alimentarius Commission). 1978. *Report of the 10th Session of the Coordinating Committee for Europe, Vienna, 13–17 June 1977*. ALINORM 78/19. FAO, Rome

Cominazzini, C. 1978. *Escherichia coli* and coliforms. In *Faecal Bacterial Indicators in Foods and Drinking Water* (in Italian). Società Italiana de Microbiologia Applicata, Milan

EEC. 1980a. *Council Directive for Natural Mineral Waters*. 80/777–15 July 1980. European Economic Community, Brussels

– 1980b. *Council Directive for Potable Waters*. 80/778–15 July 1980. European Economic Community, Brussels

FAO/WHO. 1981. *Report of the Fourth FAO/WHO Working Group on the Establishment and Application of Microbiological Criteria for Foods – Dried Milk Products and Natural Mineral Water*, Washington, DC 10–14 November 1980. (Document VPH/81, 32) WHO, Geneva

Gangarosa, E.J., Bisno, A.L., Eichner, E.R., Treger, M.D., Goldfield, M., DeWitt, W.E., Fodor, T., Fish, S.M., Dougherty, W.J., Murphy, J.B., Feldman, J., and Vogel, H. 1968. Epidemic of febrile gastroenteritis due to *Salmonella java* traced to smoked whitefish. *American Journal of Public Health, 58*, 114–121

Gangarosa, E.J., Kendrick, M.A., Loewenstein, M.S., Merson, M.H. and Mosley, J.W. 1980. Global travel and travelers health. *Aviation, Space and Environmental Medicine, 51*, 265–270

Geldreich, E.E. 1978. Bacterial populations and indicator concepts in feces sewage, storm water, and solid wastes. In *Indicators of Viruses in Water and Food*, ed. G. Berg. Ann Arbor Science Publication, Ann Arbor, Michigan

Greenberg, A.E., Conners, J.J., Jenkins, D., and Franson, M.A.H. 1981. *Standard Methods for the Examination of Water and Wastewater*, 15th ed. American Public Health Association, Washington, DC

Hoadley, A.W. 1976. Potential health hazards associated with *Pseudomonas aeruginosa* in water. In *Bacterial Indicators / Health Hazards Associated with Water*, ed. A.W. Hoadley and B.J. Dutka. American Society for Testing and Materials, Philadelphia

Hoadley, A.W., and Dutka, B.J. (eds). 1976. *Bacterial Indicators / Health Hazards Associated with Water*. American Society for Testing and Materials, Philadelphia

Holden, W.S. 1970. *Water Treatment and Examination*. Williams and Williams Company, Baltimore

Howie, J.W. 1968. Typhoid in Aberdeen, 1964. *Journal of Applied Bacteriology, 31*, 171–178

ICMSF. 1978. *Microorganisms in Foods. 1. Their Significance and Methods of Enumeration*, 2nd ed. University of Toronto Press, Toronto

– 1980a. *Microorganisms in Foods*. Vol. 1. *Factors Affecting Life and Death of Microorganisms*. Academic Press, New York

– 1980b. *Microorganisms in Foods*. Vol. 2. *Food Commodities*. Academic Press, New York

Levine, R.J., and Nalin, D.R. 1976. Cholera is primarily waterborne in Bangladesh. *Lancet, ii*, 1305

Morisetti, M.D. 1979. Microbiological standards for natural waters. *Nuovi Annali d'Igiene e Microbiologia, 30* (No. 3–4), 253–264

Mossel, D.A.A. 1976. Various taxo- and ecogroups of bacteria as index organisms for the enteric contamination of bottled waters: Their significance and enumeration. *Annali dell'Istituto Superiore di Sanità, 12*, 177–190

– 1977. Microbiological quality assurance of water in relation to food hygiene. *Archiv für Lebensmittelhygiene, 28*, 1–2

Mossel, D.A.A., and Oei, H.Y. 1975. Person-to-person transmission of enteric bacterial infection. *Lancet, i*, 751

Mossel, D.A.A., Harrewijn, G.A., and van Sprang, F.J. 1973. Microbiological quality assurance for weaning formulae. In *The Microbiological Safety of Food*, ed. B.C. Hobbs and J.H.B. Christian, 77–88. Academic Press, London

Mossel, D.A.A., de Vor, H., and Eelderink, I. 1976. A further simplified procedure for the detection of *P. aeruginosa* in often contaminated aqueous substrata. *Journal of Applied Bacteriology, 41*, 307–309

Mossel, D.A.A., van Ekeren, A.J.W.M., and Eelderink, I. 1977. A simplified procedure for the examination of drinking water for bacteria of public health significance: The differential hydrobacteriogramme. *Zentralblatt für Bakteriologie-Hygiene*, I. Abt. Orig. B *165*, 498–516

Nolan, C., Harris, N., Ballard, J., Allord, J., and Kobayashi, J. 1982. Outbreak of *Yersinia enterocolitica* – Washington State. *Morbidity and Mortality Weekly Report (U.S.), 31*, 562–564

Panero, C., Grasso, A., La Canza, C., Raguzzini, R., and Bosco, G. 1970. Pollution of drinking water and infections caused by *Pseudomonas aeruginosa* in a paediatric ward for immature babies (in Italian). *Giornale di Batteriologia, Virologia, Immunologia, 63*, 517–527

Schmidt-Lorenz, W. 1976. Microbiological characteristics of natural mineral water. *Annali dell'Istituto Superiore di Sanità, 12*, 93–112

Weber, G., Werner, H.B. and Matschnigg, H. 1977. *Pseudomonas aeruginosa* in Trinkwasser als Todesursache bei Neugeborenen. *Zentralblatt Bakteriologie Parasitenk. Infektionskr. Hygiene*, I. Abt. Orig. *216*, 210–214

WHO (World Health Organization). 1971. *International Standards for Drinking Water*, 3rd ed. WHO, Geneva

– 1979. *Human Viruses in Water, Wastewater, and Soil*. Technical Report Series 639. WHO, Geneva

– 1980. *Guidelines for Drinking Water Quality – Microbiological Aspects (Background Information)*. WHO, Geneva

Young, V.M. (ed.) 1977. *Pseudomonas aeruginosa: Ecological Aspects and Patient Colonization*. Raven Press, New York

26

Shelf-stable canned foods

Shelf stable canned foods are packed in hermetically sealed containers and are commercially sterile. Commercial sterility of thermally processed food means the condition achieved by the application of heat, alone or in combination with other treatments, to render the food free from microorganisms capable of growing in the food under normal conditions of distribution and storage. Such processes include: (a) the 'botulinum cook' used for low-acid foods (i.e., those having an equilibrium pH above 4.6); (b) the lesser heat-treatment applied to products containing curing ingredients such as sodium chloride or sodium nitrite; (c) the relatively mild heat-treatment given to products of pH 4.6 and below ('acid' foods); and (d) the mild heat-treatment given to foods in which the water activity, or combination of water activity with other preservation factors such as pH, inhibits growth from bacterial spores.

Many canned foods have the potential to cause botulism. However, *Clostridium botulinum* is present in such low numbers, and so infrequently, in such products that no realistic sampling plan would be adequate as a direct measure of its presence.

In practice the controls applied consist of in-plant control to ensure that (i) the containers are sound and correctly seamed or sealed, (ii) the correct thermal process has been applied, (iii) the cooling water is of suitable microbiological quality, either by chlorination and chemical monitoring or by microbiological monitoring, and (iv) hygienic practices during and after cooling are satisfactory.

A. RECOMMENDED HYGIENIC PRACTICES FOR CANNED FOODS

Within recent years a number of codes of practice for commercial processing of foods in hermetically sealed containers have been published. For example, the United States Food and Drug Administration has issued a code of good manufacturing practice for low-acid canned foods (FDA, 1979) as has the UK Department of Health and Social Security (DHSS, 1981). The Codex Alimentarius Commission has drafted a *Recommended International Code of Hygienic Practice for Low-Acid and Acidified Low-Acid Canned Foods* (Codex, 1983).

B. INTEGRITY OF CONTAINERS

Even when an adequate heat-process has been applied, the integrity of hermetically sealed containers used for canned foods is critical to safe processing and requires constant surveillance by the can manufacturer and canner. A defective or unsound container is one that is of faulty manufacture, is improperly closed, or has been damaged in such a manner as to permit recontamination of the can contents following the heat-process (post-processing contamination). If the contaminant is a pathogen and the food is capable of supporting its growth, a health hazard exists, as demonstrated by type E botulism from canned tuna (Johnston et al., 1963), typhoid from canned corned beef (Howie, 1968); and staphylococcal food-poisoning from canned peas (Bashford et al., 1960). Even a sound container, if mishandled, may leak, and if hygiene is poor, contamination of the contents may occur (Put et al., 1972; Stersky et al., 1980). To minimize the chances of contamination caused by defective containers, the container manufacturers, as well as the canners, should maintain a strict quality assurance program including pressure-testing and can seam measurements to ensure that predetermined tolerances are met. Minimum inspection programs for seam and seal checks are detailed in FDA, 1978, and in guidelines issued by the NFPA (1979). More detailed guidance on can inspection is available in HPB, 1983; OSU 1982; and Thorpe and Barker, 1984, which also includes information on containers with welded side seams.

C. COOLING WATER

In addition to container-integrity checks and process controls, the canner should use water of suitable microbiological quality for cooling the retorted cans. If appropriate, the water should be chlorinated or otherwise sanitized. Examination of the cooling water is the responsibility of the canner and should involve regular microbiological examination and, if a sanitizer is used, frequent analysis for the concentration of the sanitizer present.

D. INCUBATION TESTS

Opinions differ concerning the value of incubating and testing a small number of cans immediately after processing, but there is no doubt that such incubation tests are of no value in ensuring the safety of shelf-stable canned foods. It must also be borne in mind that such limited incubation tests will reveal only gross underprocessing and/or extensive post-processing contamination. However, limited incubation tests are still required by some agencies for certain low-acid canned foods which do not receive a full botulinum cook, including some cured meats which are shelf-stable as a result of the presence of curing agents. Incubation tests may also be of use to the manufacturer to monitor spoilage trends over long periods of time. It is accepted practice to examine all cans subjected to incubation tests for swelling, but it must be realized that a variable proportion of contaminated cans produce swells. Hence, the contents of as many of the incubated cans as practicable should be examined for pH change and other evidence of spoilage. The practice of deliberate storage (or incubation) of all cans after thermal processing, and removal of swollen cans prior to labelling or distribution, is to be discouraged. It can lead to the marketing of an unsatisfactorily processed product, which may be a health hazard as some pathogens (e.g., staphylococci and *Salmonella typhi*) do not produce gas in many canned products. It should be emphasized that no sampling procedure ensures adequate safety with respect to the botulism hazard. Industrial experience suggests that greater reliance should be placed on the following: thermal-processing records, constancy of product formula (solid-to-liquid ratios), volume of fill, the use of automated instrumentation such as time-temperature recording devices, and alarm warning systems to signal any interruption of the specified thermal process.

E. PROCESS VALIDATION AND RECORDS

In addition to data on seam inspection (e.g., tear-down and visual inspection), thermal processing, and cooling-water quality, it is recommended that information on appropriate chemical composition (e.g., salt and nitrite content), pH, or any other critical factors be recorded and made available to the controlling agency. The data on chemical composition should give, at least, mean values and the standard deviation or range. If this information gives cause for concern, or is lacking or inadequate, further can examinations appear advisable (see Chapter 5, Section N: Investigational Sampling).

F. SAMPLING

For reasons stated above, a sampling plan for the routine surveillance of shelf-stable canned foods to determine microbiological safety is not recommended.

REFERENCES

Bashford, T.E., Gillespie, T.G. and Tomlinson, A.J.H. 1960. *Staphylococcal Food Poisoning Associated with Processed Peas*. Campden Food Preservation Research Association, Chipping Campden, Gloucestershire, England

Codex (Codex Alimentarius Commission). 1983. *Recommended International Code of Practice for Low Acid and Acidified Low-Acid Canned Foods*, 1st ed. CAC/RCP 23–1979, FAO, Rome

DHSS (Department of Health and Social Security). 1981. *Code of Hygienic Practice for Low-Acid Canned Foods*, HMSO, London

FDA (United States Food and Drug Administration). 1978. *Bacteriological Analytical Manual*, 5th ed. Form FD2905 (4/76), Chapter 22. Examination of metal container integrity (R.C. Lin, P.H. King, and M.R. Johnston)

– 1979. *Rules and Regulations 21CFR Part 113 – Thermally Processed Low-Acid Foods Packaged in Hermetically Sealed Containers Current Good Manufacturing Practice*. Federal Register, 44. No. 53 16209–16238

Howie, J.W. 1968. Typhoid in Aberdeen 1964. *Journal of Applied Bacteriology, 31*, 171–178

HPB (Health Protection Branch). 1983. Draft Document MFA–25. *Methods for the Evaluation of Canned Foods and Their Containers*. HPB, Health and Welfare Canada, Ottawa

Johnston, R.W., Feldman, J., and Sullivan, R. 1963. Botulism in canned tuna fish. *Public Health Reports*, 78, 561–564

NFPA (National Food Processors Association). 1979. *Guidelines for Evaluation and Disposition of Damaged Canned Food Containers*, 2nd ed. Bulletin 38L, NFPA, Washington, DC

OSU (Oregon State University, Extension Service of the Department of Food Science and Technology). 1982. *Top Double Seaming Manual*, 5th printing. OSU, Corvallis, Oregon

Put, H.M.C., Van Doren, H., Warner, W.R., and Kruiswijk, J.T.H. 1972. The mechanism of microbiological leaker spoilage of canned foods: A review. *Journal of Applied Bacteriology, 35*, 7–27

Stersky, A., Todd, E., and Pivnick, H. 1980. Food poisoning associated with post-process leakage (P.P.L.) in canned foods. *Journal of Food Protection*, *43*, 465–467

Thorpe, R.H., and Barker, P.M. 1984. *Visual Can Defects*. The Campden Food Preservation Research Association, Chipping Campden, Gloucestershire, England

APPENDICES

Appendix 1

OBJECTIVES AND ACCOMPLISHMENTS OF THE INTERNATIONAL COMMISSION ON MICROBIOLOGICAL SPECIFICATIONS FOR FOODS

HISTORY AND PURPOSE

The International Commission on Microbiological Specifications for Foods (ICMSF) was formed in 1962 by the parent body, the International Association of Microbiological Societies (IAMS) (now the International Union of Microbiological Societies [IUMS]), in response to the need for internationally acceptable and authoritative decisions on microbiological limits for foods commensurate with public health safety, and particularly for foods moving in international commerce. Its overall purpose is to appraise public health aspects of the microbiological content of foods. Through the IUMS, the ICMSF is linked to the International Union of Biological Societies (IUBS) and to the World Health Organization (WHO) and, hence, is a body of the United Nations.

The founding terms of reference are as follows: (a) to assemble, correlate, and evaluate evidence about the microbiological quality of foods; (b) to consider whether microbiological criteria are necessary for any particular food; (c) where necessary, to propose such criteria; and (d) to suggest appropriate methods of sampling and examination. More descriptively, the ICMSF seeks to provide comparable standards of judgment in different countries, to foster safe movement of foods in international commerce, and to dissipate difficulties caused by disparate microbiological criteria and methods of analysis. Fulfilment of such objectives would be of great value to the food industry, to the expansion of international trade in foods, to national control agencies, to the international agencies more concerned with the humanitarian aspects of food distribution, and, eventually, to the health of the consuming public.

The ICMSF is essentially a scientific advisory body that provides basic information through extensive study and makes recommendations based on such information. The results of studies are published as either books or papers (see Appendix 4) and are thus available to interested individuals, governments, and national and international organizations to use as desired. The group provides the facts without prejudice and thereby fills a useful role as an authoritative base. Primarily through cross-membership, close liaison is enjoyed with other organizations involved in international standards, such as the Codex Alimentarius Commission, the International Standards Organization, the International Dairy Federation, and the Association of Official Analytical Chemists.

At meetings, the ICMSF functions as a work party, not as a forum for the reading of papers. Much of the work is done by subcommittees during the interval between meetings, often with assistance of non-member consultants. The general meetings are largely directed toward assessing the work of the subcommittees, debating to achieve a consensus, editing of draft submissions, and planning. Meetings have been held in Montreal, Canada (1962, the founding meeting); Cambridge, England (1965); Moscow, USSR (1966); London, England (1967); Dubrovnik, Yugoslavia (1969); Mexico City, Mexico (1970); Opatija, Yugoslavia (1971); Langford, England (1972); Ottawa, Canada (1973); Caracas, Venezuela (1974); Alexandria, Egypt (1976); Cairo, Egypt (1977 and 1978); Stresa, Italy (1980), Chexbres, Switzerland (1981); Anaheim, California, USA (1982), Sharnbrook, Bedford, England (1983); and Berlin (West), Federal Republic of Germany (1984).

MEMBERSHIP AND SUBCOMMISSIONS

The membership consists of 19 food microbiologists from 11 countries (see listing in Appendix 2), whose combined professional interests include research, public health, official food control, education, and industrial research and development. They are drawn from governmental laboratories in public health, agriculture, and food technology, from universities, and from the food industry. In addition, the ICMSF engages consultants from time to time to assist with specific aspects of its studies. All members and consultants are chosen on the basis of their expertise in areas of food microbiology, not as national delegates; all work is voluntarily without fees or honoraria.

To promote similar activities among food microbiologists on a regional scale, subcommissions have been created in various areas of the world. To date, three have been established (see Appendix 2): one in the Balkan-Danubian region (the Balkan-Danubian Subcommission, BDS), composed of six members; one in Latin America (the Latin-American Subcommission, LAS), composed of 11 members; and one in the Middle East–North African region (the Middle East–North African Subcommission, MENAS), composed of eight members. Each is an autonomous body which conducts studies on problems of specific concern to its region and holds meetings.

METHODS-TESTING PROGRAM

The overall objective is to determine by detailed comparative analysis, involving laboratories in various countries, the best methods for the enumeration and identification of indicator and food-poisoning bacteria. In *Microorganisms in Foods 1: Their Significance and Methods of Enumeration*, the ICMSF describes several of the best-known methods for some of the microbial categories, because it could not distinguish which one, if any, is superior. International comparative testing is seen as the only way to determine the most accurate methods.

The studies have been coordinated by Silliker Laboratories, Chicago Heights, Illinois, in the United States (Drs J.H. Silliker and D.A. Gabis, coordinators), and by the Health Protection Branch of Health and Welfare Canada, in Ottawa, Canada

(Drs H. Pivnick and K. Rayman, coordinators). All projects are planned by a committee of ICMSF members and consultants and are approved by the ICMSF in plenary session. Studies completed to date or in progress include those on methods for *Salmonella*, coliforms, *Staphylococcus, Clostridium perfringens, Escherichia coli,* and *Vibrio parahaemolyticus* (see Appendix 4).

FINANCING

The ICMSF raises funds for its activities from government agencies in several countries, from the WHO, from the IUBS, and from the food industry. Assistance from governmental agencies has come in the form of grants for specific projects: the United States Department of Agriculture has given two grants in support of the methods-testing program; the United States Department of Health, Education and Welfare sponsored PL 480 grants to support two general meetings on sampling of foods and to support three others on the preparation of *Microbial Ecology of Foods*; the government of Kuwait also contributed financially to the preparation of *Microbial Ecology of Foods*; and Health and Welfare Canada supported one general meeting of the preparation of the 2nd edition of *Microorganisms in Foods, 1: Their Significance and Methods of Enumeration.* Several groups in Germany sponsored a planning meeting for a new book on Control of Operations for Assurance of Safety and Quality. WHO has contributed annually in support of the methods-testing program and general expenditures, and the IUBS has granted funds in support of administration costs and to help meet expenses for the publication of this book. Over 80 food companies and agencies in 11 countries contribute to the ICMSF Sustaining Fund, mostly on a yearly basis (see Appendix 3).

Appendix 2

MEMBERS OF THE ICMSF AND ITS SUBCOMMISSIONS

MEMBERS OF THE ICMSF

Dr J.H.B. Christian (*Chairman*), Chief, Division of Food Research, CSIRO, P.O. Box 52, North Ryde, NSW 2113, Australia

Dr Frank L. Bryan (*Secretary*), 2022 LaVista Circle, Tucker, Georgia, USA 30084 (Formerly Scientist Director, Centers for Disease Control, US Public Health Service, Atlanta, Georgia, USA)

Dr J.H. Silliker (*Treasurer*), President, Silliker Laboratories, 1139 E. Dominguez Street, Suite I, Carson, California, USA 90746

Dr A.C. Baird-Parker, Head, Microbiological and Preservation Research, Unilever Research, Colworth Laboratory, Colworth House, Sharnbrook, Bedford, MK 44 1LQ, England

Dr F.F. Busta, Chairman, Food Science and Human Nutrition Department, University of Florida, 359 Food Science Department, Gainesville, Florida, USA, 32611

Dr J.M. Goepfert, Associate Director of Research, Canada Packers Inc., 2211 St. Clair Ave. W., Toronto, Ontario, Canada, M6N 1K4

Dr F.H. Grau, CSIRO, Division of Food Research, Meat Research Laboratory, P.O. Box 12, Canon Hill, QLD 4170, Australia

Dr Betty C. Hobbs, formerly Director, Food Hygiene Reference Laboratory, Central Public Health Laboratory, Colindale Avenue, London, NW9 5HT, England

Dr Milica Kalember-Radosavljevic, Food Bacteriologist, Military Medical Academy, Institute of Hygiene, 2 Pasterova Avenue, Belgrade, Yugoslavia

Professor J. Liston, Director, Institute for Food Science and Technology, HF–10 (School of Fisheries, College of Ocean and Fisheries Science), University of Washington, 3707 Brooklyn Avenue N.E., Seattle, Washington, USA 98195

Dr P.M. Nottingham, Head, Science Division, The Meat Industry Research Institute of New Zealand (Inc.), P.O. Box 617, Hamilton, New Zealand

Dr F. Quevedo, Head, Food Protection, Pan American Health Organization, 525 Twenty-third Street N.W., Washington, DC., USA, 20037 (Formerly Chief, Laboratory Services and Head, Food Microbiology and Hygiene Unit, Pan American Zoonoses Centre)

Dr T.A. Roberts, Head, Microbiology Division, Agricultural and Food Research Council, Food Research Institute – Bristol, Langford, Bristol, BS18 7DY, England

Dr A.N. Sharpe, Head, Microbiology Automation Branch, Health and Welfare Canada, Tunney's Pasture, Ottawa, Ontario, Canada K1A OL2

Mr B. Simonsen, Head of Division, Danish Meat Products Laboratory, Ministry of Agriculture, Howitzvej 13, DK-2000, Copenhagen F, Denmark

Dr H.J. Sinell, Director, Institute of Food Hygiene, Free University of Berlin, 1000 Berlin (West) 33 (Dahlem), Koserstr. 20, Germany

Dr Paul Teufel, Institut für Veterinärmedizin des Bundesgesundheitsamtes (Robert von Ostertag-Institut) Bundesgesundheitsamt, Postfach 330013 (West), D-I000 Berlin (West)-33

Dr R. Bruce Tompkin, Chief Microbiologist, Beatrice Refrigerated Food Group, 1919 Swift Drive, Oakbrook, Illinois, USA, 60521

Dr M. van Schothorst, Central Quality Assurance Laboratory, Nestec, Case postale 88, 1814 La Tour de Peilz, Switzerland

PAST MEMBERS OF THE ICMSF

Dr M.T. Bartram, USA, Member 1967–68
Dr H.E. Bauman, USA, Member 1964–77
Dr Leon Buchbinder, USA, Member 1962–65 (Deceased)
Dr R. Buttiaux, France, Member 1962–67 (Deceased)
Dr D.S. Clark, Canada, Member 1963–85
Dr C. Cominazzini, Italy, Member 1962–83
Dr C.E. Dolman, Canada, Member 1962–73
Dr R. Paul Elliott, USA, Member 1962–77 (Deceased)
Professor O. Emberger, Czechoslovakia, Member 1971–86 (Deceased)
Mrs Mildred Galton, USA, Member 1963–68 (Deceased)
Dr E.J. Gangarosa, USA, Member 1968–70
Dr H.E. Goresline, USA/Austria, Member 1963–70
Dr A. Hurst, United Kingdom/Canada, Member 1963–69
Dr H. Iida, Japan, Member 1966–77
Dr M. Ingram, United Kingdom, Member 1968–74, Ex-officio member 1962–68 (Deceased)
Dr K.H. Lewis, USA, Member 1962–82
Dr Holger Lundbeck, Sweden, Member 1962–83, Chairman 1973–80
Dr G. Mocquot, France, Member 1962–80
Dr G.K. Morris, USA, Member 1970–74
Dr D.A.A. Mossel, Netherlands, Member 1962–75
Dr N.P. Nefedjeva, USSR, Member 1963–79
Dr C.F. Niven, Jr, USA, Member 1974–81
Dr J.C. Olson, Jr, USA, Member 1968–82
Dr H. Pivnick, Canada, Member 1974–83
Dr G.G. Slocum, USA, Member 1962–68 (Deceased)
Dr F.S. Thatcher, Canada, Member and Chairman 1962–73

MEMBERS OF THE ICMSF SUBCOMMISSIONS

Balkan and Danubian Subcommission

Dr Vladimir Bartl (*Chairman*), Head, Hygiene Laboratories, Hygiene Station for Middle Czech Region, Safarikova 14, 120 00 Praha 2, Czechoslovakia

Dr HAJNALKA DOMJAN KOVACS (*Secretary/Treasurer*), Central Laboratory, Hygienic Control of Food Industry of Ministry of Agriculture and Food, 1453 Budapest P.O.B. 13, Hungary/Mester U.81

Dr MILICA KALEMBER-RADOSAVLJEVIC, Food Bacteriologist, Military Medical Academy, Institute of Hygiene, 2 Pasterova Avenue, Belgrade, Yugoslavia

Professor Dr O. PRANDL, Director, Institute of Meat Hygiene and Veterinary Food Technology, Vienna III/40, linke Bahngasse 11, Austria

Dr S. TZANNETIS, Faculty of Medicine, Department of Microbiology, National University of Athens, Athens 609, P.O. Box 1540, Greece

Professor Dr C. ZACHARIEV, Director, N.P.O. Veterinarno Delo, U.L. Rabotniceska Klasa No. 1 Sofia, Bulgaria

Latin-American Subcommission

Professora JOSEFINA GOMEZ-RUIZ (*Chairwoman*), Apartado 50259, Caracas 1050 A, Venezuela

Dra SYLVIA MENDOZA G. (*Secretary-Treasurer*), Division of Biological Sciences, Department of Bioengineering, Simon Bolivar University, Apartado 80659, Caracas 108, Venezuela

Dr HERNAN PUERTA CARDONA, Chairman, Food Hygiene Section, Escuela Nacional de Salud Public, Universidad de Antioquia, Apartado Aereo 51922, Medellin, Colombia

Professora NENUFAR SOSA DE CARUSO, Director, Dairy Institute, Veterinary Faculty, University of Uruguay, Casilla de Correo 753, Montevideo, Uruguay

Dra YOLANDA ORTEGA DE GUTIERREZ, Lab. Nacional de Referencia, Calz. de Tlalpan 4492, Ca. Toriello Guerra Cep:14050, Mexico D.F.

Dr SEBASTIAO TIMO IARIA, Instituto de Ciencias Biomedicas, Universidade de Sao Paulo, Av. Dr. Arnaldo 715–C.P. 8099, Sao Paulo, E. Sao Paulo, Brazil

Dr MAURO FABER DE FREITAS LEITAO, Head, Department of Food Microbiology, Instituto de Technologia de Alimentos, Caixa Postal 139, 13.100 Campinas Sao Paulo, Brazil

Dra ELIANA MARAMBIO, Instituto de Salud Publica, Casilla 48, Santiago, Chile

Dr FERNANDO QUEVEDO, Head, Food Protection, Pan American Health Organization, 525 Twenty-third Street N.W., Washington, DC, USA, 20037 (Formerly Chief, Laboratory Services and Head, Food Microbiology and Hygiene Unit, Pan American Zoonoses Centre)

Dra MARIA ALINA RATTO, Head, Centro Latinoamericano, de Ensenanza e Investigacion de Bacteriologia Alimentaria (CLEIBA), Universidad Nacional Mayor de San Marcos, Apartado 5653, Lima, Peru

Dra ELVIRA REGUS, C.N. Penson, 116, Apt 1 – 1 Edificio Paraiso V., Santo Domingo, Dominican Republic

Middle East–North African Subcommission

Professor REFAT HABLAS (*Chairman*), Bacteriological Department, Faculty of Medicine, Al-Azhar University, El Houssein Hospital, Eldarrasa, Cairo, Egypt.

Dr HASSAN SID AHMED (*Secretary*), Head, Department of Bacteriology, P.O. Box 287, National Health Laboratory, Khartoum, Sudan

Professor A. ALAOUI, Director of Institute Pasteur Maroc, Professor of Microbiology, Casa School of Medicine, Casablanca, Morocco

Dr ABDUL-KAREEM NASIR AL-DULAIMI, Head, Department of Food and Water Hygiene, Central Public Health Laboratories, Andulis Square, Alwia, Baghdad, Iraq

Mr I. KASHOULIS, Analyst, Government Laboratory, Ministry of Health, Nicosia, Cyprus

Professor EL-SAYED EL-MOSSALAMI, Head, Meat Hygiene Department, Faculty of Veterinary Medicine, Cairo University, Giza, Egypt

Mr YACOUB KHALID MOTAWA, Head of Food Control Laboratory, Microbiological Section, Preventive Health, Public Health Laboratory, Ministry of Health, Kuwait

Dr NEJI OTHMAN, Chief, Laboratory of Food Microbiology, Ministry of Health, National Institute of Nutrition, Tunis, Tunisia

CONSULTANTS FOR THIS REVISION

G. JARVIS, Health and Welfare Canada, Health Protection Branch, 4th Floor, Tower B, 355 River Road, Vanier, Ontario, Canada K1A 1B8

D. KILSBY, Microbiological and Preservation Research, Unilever Research, Colworth Laboratory, Colworth House, Sharnbrook, Bedford, MK44 1LQ, England

Appendix 3

CONTRIBUTORS TO THE ICMSF SUSTAINING FUND FOR 1970–1984

Alivar Foundation, Italy

Amatil Ltd., Box 145 GPO Sydney, NSW 2001, Australia

American Can Co., American Lane, Greenwich, Connecticut, USA, 06830

Arnotts Biscuits Co., Ltd., 170 Kent Street, Sydney, NSW 2000, Australia

Atlantic Sugar Ltd., P.O. Box 7, Montreal, Quebec, Canada, H3C 1C5

Bacon and Meat Manufacturers Association, 1–2 Castle Lane, London, SW1 E 6DU, England

Banca Popolare di Novara, Novara, Italy

Beatrice Foods Co., 120 South La Salle, Chicago, Illinois, USA, 60603

Beecham Group Ltd., Beecham House, Great West Road, Brentford, Middlesex, England

Brooke-Bond Liebig Ltd., European Technical Division, Trojan Way, Purley Way, Croydon CR9 9EH, England

Brown and Polson Ltd., Clay Gate House, Littleworth Road, Esher, Surrey, England

Burns Food Ltd., P.O. Box 1300, Calgary, Alberta, Canada, T2P 2L4

Cadbury Schweppes Co., Ltd., P.O. Box 88, St. Kilda West, Victoria 3182, Australia

Cadbury Schweppes Powell Ltd., 1245 Sherbrooke Street W., Suite 1625, Montreal, Quebec, Canada, H3G 1G6

Campbell Institute for Research and Technology, Campbell Place, Camden, New Jersey, USA, 08101

Campbell Soup Co. Ltd., 60 Birmingham Street, Toronto, Ontario, Canada, M8V 2B8

Canada Packers Ltd., 2211 St. Clair Avenue W., Toronto, Ontario, Canada, M6N 1K4

Carlo Erba Institute for Therapeutic Research, Milan 20159, Italy

Central Alberta Dairy Pool, 5302 Gaetz Avenue, Red Deer, Alberta, Canada

Centro Studi sull'Allimentazione, Gino Alfonso Sada, P.za Diaz 7–20123, Milan, Italy

Christie Brown and Company Ltd., 2150 Lakeshore Boulevard W., Toronto, Ontario, Canada, M8V 1A3

Coca-Cola Export Corporation, 260 Peachtree Street N.W., Atlanta, Georgia, USA 30303
Comet Confectionery Limited, P.O. Box 398, 2950 Nelson St., St Hyacinthe, Quebec, Canada, J2S 1Y7
CPC International, Inc., International Plaza, Englewood Cliffs, New Jersey, USA 07632
CSR Ltd., Box 1630 GPO, Sydney, NSW 2001, Australia
Del Monte Corporation, 215 Fremont Street, San Francisco, California, USA 94119
Difco Italia, Italy
Difco Laboratories, Detroit, Michigan, USA 48232
Distillers Company, Ltd., 21 St. James Square, London, SW 1, England
Export Packers Company Ltd., 250 Summerlea Road, Bramalea, Ontario, Canada, L6T 3V6
Findus Ltd., Bjuv, Sweden
Frigoscandia Ltd., Fack S-215 ol, Helsingborg, Sweden
Gelda Scientific and Industrial Development, Inc., Mississauga, Ontario, Canada
General Foods Canada, Ltd., 520 William St., Cobourg, Ontario, Canada, K9A 4L4
General Foods Corporation, Technical Center, White Plains, New York, USA 10602
Gerber Products Co., 445 State Street, Fremont, Michigan, USA 49412
H.J. Heinz Company Ltd., Hayes Park, Hayes, Middlesex, England
H.J. Heinz Company Australia Limited, P.O. Box 37, Dandenong, Victoria, 3175, Australia
Home Juice Company, Ltd., 175 Fenmar Drive, Weston, Ontario, Canada
Horne and Pitfield Foods Ltd., 14550 – 112th Avenue, P.O. Box 2266, Edmonton, Alberta, Canada
Indulac, Caracas, Venezuela
Infant Formula Council, 64 Perimeter Center East, Atlanta, Georgia, USA 30346
ITT Continental Baking Company, P.O. Box 731, Rye, New York, USA 10580
IUMS, 31, Chemin Joseph-Aiguier, B.P. 71 – 13277 Marseille Cedex 9, France
J. Sainsbury Ltd., Stamford House, Stamford Street, London SE1 9LC, England
John Labatt Ltd., 451 Ridout Street, London, Ontario, Canada, N6A 2P6
Joseph Rank Ltd., Millcrat House, Eastcheap, London, EC 3, England
Kellogg/Salada Canada Ltd., 6700 Finch Avenue W., Rexdale, Ontario, Canada, M9W 5P2
Kraft Foods Ltd., Box 1673N, BPO, Melbourne, 3001, Australia
Kuwait Ministry of Health, Kuwait, Kuwait
Lahgese-Iglo Gmbh, Hauptverwaltung, Postfach 10 40 29, 2000 Hamburg 1, West Germany
Maizena Gesellschaft mbH, Postfach 560, Knorrstr. 1, 71 Heilbronn/Neckar, Germany
Maple Leaf Mills Ltd., P.O. Box 370, Station A, Toronto, Ontario, Canada, M5W 1C7
Marks and Spencer (UK) Ltd., Michael House, Baker Street, London W.1, England
Mars (UK) Ltd., Dundee Road, Slough SL1 4JA, England
McCormick and Company, Inc., Baltimore, Maryland, USA 21202
Milk Marketing Board, Thames Ditton, Surrey, England
Nabisco Brands Ltd., 1 Dundas St. W., Toronto, Ontario, Canada, M5G 2A9

Nestlé Products Technical Assistance Co., Ltd., Case postale 88, CH–1814 La Tour de Peilz, Switzerland

New Zealand Food Manufacturers' Federation, Industry House, 38–44 Courtenay Place, Wellington, New Zealand

Oscar Mayer Company, Caracas, Venezuela

Plasmon SPA, Corso, Garibaldi 97.99, 20121 Milan, Italy

Provincia Novara, Novara, Italy

RHM Research Ltd., Lincoln Road, High Wycombe, Bucks., H12 3QM, England

RJR Foods Inc., 4th and Main Streets, Winston-Salem, North Carolina, USA 27102

Reckitt and Coleman Ltd., Carrow, Norwich NR1 2DD, England

Regione Lombardia, Italy

Regione Piemonte, Italy

Ross Laboratories, 625 Cleveland Avenue, Columbus, Ohio, USA 43216

Royal Company, Caracas, Venezuela

Silliker Laboratories, 1304 Halsted Street, Chicago Height, Illinois, USA 60411

Spillers Ltd., Old Charge House, Cannon Street, London EC 4, England

Standard Brands Canada Ltd., 31 Airlie Street, LaSalle, Quebec, Canada

Swift Canadian Company, Ltd., 2 Eva Road, Etobicoke, Ontario, Canada, M9C 4V5

Tate and Lyle Ltd., Philip Lyle Memorial Research Laboratory, P.O. Box 68, Reading RG6 2BX, England

Terme de Crodo, via Cristoforo Gluck 35, Milan 20125, Italy

Tesco Stores Ltd., Tesco House, Delamere Road, Cheshunt, Waltham Cross, Herts., England

The Borden Company Ltd., 1275 Lawrence Avenue E., Don Mills (Toronto), Ontario, Canada

The J. Lyons and Company Ltd., Cadby Hall, London W14 09A, England

The Nestlé Company (Australia) Limited, Box 4320, GPO, Sydney, NSW 2001, Australia

The Pillsbury Company, 608 Avenue S., Minneapolis, Minnesota, USA 55402

The Quaker Oats Company, 617 West Main Street, Barrington, Illinois, USA 60010

The Quaker Oats Company, Canada Ltd., Quaker Park, Peterborough, Ontario, Canada

Thomas J. Lipton, Inc., 800 Sylvan Avenue, Englewood Cliffs, New Jersey, USA 07632

Thomas J. Lipton, Ltd., 2180 Yonge Street, Toronto, Ontario, Canada, M4S 2C4

TQC Congress (SINU), Italy

Unigate Australia Company Ltd., P.O. Box 13, Dandenong, Victoria 3175, Australia

Unilever Ltd., Unilever House, Blackfriars, London, EC 4, England

Wattie Industries Ltd., P.O. Box 439, Hastings, New Zealand

World Health Organization, Geneva, Switzerland

Appendix 4

PUBLICATIONS OF THE ICMSF

BOOKS

1 *Microorganisms in Foods, 1. Their Significance and Methods of Enumeration*, 2nd ed. 1978. Author ICMSF. University of Toronto Press, Toronto, Canada M5S 1A6. ISBN 0-8020-2293-6. 434 pages (1st ed., ed. F.S. Thatcher and D.S. Clark, 1968)

2 *Microbiological Specifications and Testing Methods for Irradiated Foods*. 1970. Compiled and edited in cooperation with the Food and Agriculture Organization and the International Atomic Energy Agency. Published as Technical Series 104 by the International Atomic Energy Agency, Vienna. 122 pages. Available in English, French, German, and Russian from the United Nations Bookshop at United Nations Headquarters, New York and the Publishing Section, International Atomic Energy Agency, Karntner Ring 11, P.O. Box 590, A-10I1 Vienna, Austria

3 *Microorganisms in Foods, 2: Sampling for Microbiological Analysis: Principles and Specific Applications*. 1974. Author ICMSF. University of Toronto Press, Toronto, Canada M5S 1A6. ISBN 0-8020-2143-3. 213 pages (Reprinted, with corrections, 1978)

4 *Microbial Ecology of Foods*. Volume 1: *Factors Affecting Life and Death of Microorganisms*. 1980. Author ICMSF. Academic Press, 111 Fifth Avenue, New York, New York, USA 10003. ISBN 0-12-363501-2. 332 pages (Available in soft cover)

5 *Microbial Ecology of Foods*. Volume 2: *Food Commodities*. 1980. Author ICMSF. Academic Press, 111 Fifth Avenue, New York, New York, USA 10003. ISBN 0-12-363502-0. 664 pages (Available in soft cover)

PAPERS

1 ICMSF methods studies: I. Comparison of analytical schemes for detection of *Salmonella* in dried foods. 1973. J.H. Silliker and D.A. Gabis. *Can.J. Microbiol*. 19:475–479

2 ICMSF methods studies: II. Comparison of analytical schemes for detection of

Salmonella in high moisture foods. 1974. D.A. Gabis and J.H. Silliker. *Can. J. Microbiol.* 20:663–669

3 ICMSF methods studies: III. An appraisal of 16 contemporary methods for the detection of *Salmonella* in meringue powder. 1974. E. Idziak and I.E. Erdman. *Can. J. Microbiol.* 20:703–714

4 ICMSF methods studies: IV. International collaborative assay for the detection of *Salmonella* in raw meats. 1973. I.E. Erdman. *Can. J. Microbiol.* 20:715–720

5 ICMSF methods studies: V. The influence of selective enrichment media and incubation temperatures on the detection of salmonellae in frozen meats. 1974. J.H. Silliker and D.A. Gabis. *Can. J. Microbiol.* 20:813–816

6 ICMSF methods studies: VI. The Influence of selective enrichment media and incubation temperatures on the detection of salmonellae in dried foods and feeds. 1974. D.A. Gabis and J.H. Silliker. *Can. J. Microbiol.* 20:1509–1511

7 ICMSF methods studies: VII. Indicator tests as substitutes for direct testing of dried foods and feeds for *Salmonella*. 1976. J.H. Silliker and D.A. Gabis. *Can. J. Microbiol.* 22:971–974

8 ICMSF methods studies: VIII. Collaborative study for the enumeration of *Clostridium perfringens* in foods. 1977. A.H.W. Hauschild, R.J. Gilbert, S.M. Harmon, M.F. O'Keefe, and R. Vahlefeld. *Can. J. Microbiol.* 23:884–892

9 ICMSF methods studies: IX. The Influence of selective enrichment broths, differential plating media and incubation temperatures on the detection of *Salmonella* in dried foods and feed ingredients. 1977. D.A. Gabis and J.H. Silliker. *Can. J. Microbiol.* 23:1225–1231

10 ICMSF methods studies: X. An international comparative study of four media for the enumeration of *Staphylococcus aureus* in foods. 1978. M.K. Rayman, J.J. Devoyed, U. Purvis, D. Kusch, J. Lanier, R.J. Gilbert, D.G. Till, and G.A. Jarvis. *Can. J. Microbiol.* 24:274–281

11 ICMSF methods studies: XI. Collaborative/comparative studies on determination of coliforms using the most probable number procedure. 1979. J.H. Silliker, D.A. Gabis, and A. May. *J. Food Protect.* 42:638–644

12 ICMSF methods studies: XII. Comparative study for the enumeration of *Clostridium perfringens* in feces. 1979. A.H.W. Hauschild, P. Desmarchelier, R.J. Gilbert, S.M. Harmon, and R. Vahlefeld. *Can. J. Microbiol.* 25:953–963

13 ICMSF methods studies: XIII. An international comparative study of the MPN procedure and the Anderson-Baird Parker direct plating method for the enumeration of *Escherichia coli* biotype I in raw meats. 1979. M.K. Rayman, G.A. Jarvis, C.M. Davidson, S. Long, J.M. Allen, T. Tong, P. Dodsworth, S. McLaughlin, S. Greenberg, B.G. Shaw, H.J. Beckers, S. Quist, P.M. Nottingham, and B.J. Stewart. *Can. J. Microbiol.* 25:1321–1327

14 ICMSF methods studies: XIV. Comparative study on recovery of *Salmonella* from refrigerated preenriched and enrichment broth cultures. 1983. J.-Y. D'Aoust, H.J. Beckers, M. Boothroyd, A. Mates, C.R. McKee, A.B. Moran, G.E. Sado, G.E. Spain, N.H. Sperber, A. Vassilliadis, D.E. Wagner and C. Wiberg. *J. Food Protect.* 46:391–399

15 The microbiology of specific frozen foods in relation to public health: Report of an international committee. 1963. F.S. Thatcher, *J. Appl. Bacteriol.* 26:266–285

ABOUT THE ICMSF

1 The Latin American Subcommittee on Microbiological Standards and Specifications for Foods. 1968. C.D. Saa. *Revista de la Facultad de Quimica y Farmacia* 7:8
2 El-Comite Internacional de Especificaciones Microbiologicas de los Alimentos de la IAMS. 1969. C.D. Saa. *Revista de la Facultad de Quimica y Farmacia* 8:6
3 The International Committee on Microbiological Specifications for Foods and its contribution to the maintenance of food hygiene (in Italian). 1969. C. Cominazzini. *Cronache Chimico* 25:16
4 Comision Iternacional de Especificaciones Microbiologicas de los Alimentos. 1971. S. Mendoza and F. Quevedo. *Boletin del Instituto Bacteriologico de Chili* 13:45
5 The International Committee on Microbiological Specifications for Foods. Its purposes and accomplishments. 1971. F.S. Thatcher. *J. AOAC* 54:836–841
6 The International Commission on Microbiological Specifications for Foods. 1977. D.S. Clark, *Food Technol.* 32:51–54, 67
7 International perspectives for microbiological sampling and testing of foods. 1982. D.S. Clark, *J. Food Protect.* 45:667–671
8 International Commission on Microbiological Specifications for Foods. 1984. Anon. *Food Laboratory Newsletter*, No. 1.:23–25 (Box 622, S-751 26 Uppsala, Sweden)

TRANSLATIONS

1 *Microorganisms in Foods 1. Their Significance and Methods of Enumeration.* 1st ed. 1968. Editors, F.S. Thatcher and D.S. Clark. University of Toronto Press, Toronto. (Spanish translation 1973. *Analisis Microbiologico de los Alimentos.* Editorial Acribia, Zaragoza, Spain. Dr B. Moreno Garcia, translator) No translations of the second edition are available to date.
2 *Microorganisms in Foods 2. Sampling for Microbiological Analysis: Principles and Specific Applications.* 1974. Author ICMSF. University of Toronto Press, Toronto. (Spanish translation 1981. *Micro-organismos de los Alimentos 2: Metodos de Muestreo para Analisis Microbiologicos: Principos y Aplicaciones Especificas.* Editorial Acribia, Zaragoza, Spain. Prof. Dr Juan Antonio Ordonez Pereda and Marco Antonio Diaz Hemmandez, translators)
3 *Microbial Ecology of Foods.* Vol. 1. *Factors Affecting Life and Death of microorganisms.* 1980. Author ICMSF. Academic Press, New York. (Spanish translation 1983. *Ecologia Microbiana de los Alimentos 1. Factores que afectan a la supervivencia de los microorganismos en los alimentos.* Editorial Acribia, Zaragoza, Spain. Dr D. Justino Burgos Gonzalez, Dr D. Fernando Perez Florez, Dr D. Juan Antonio Ordonez Pereda, Dra Carmen San Jose Serran, Dr D. Andes Marcos Barrado, Dr D. Francisco Sala Trepat, Dr D. Bernabe Sanz Perez, Dr D. Pascual Lopez Lorenzo, Dr D. Jose Tormo Iguacel, translators)

WHO PUBLICATIONS AUTHORED BY ICMSF

1 *Report of the WHO/ICMSF Meeting on Hazard Analysis: Critical Control Point System in Food Hygiene.* 1982. Authors: J.H. Silliker, Chairman; A.C. Baird-

Parker, F.L. Bryan, J.C. Olson, Jr., B. Simonsen, M. van Schothorst (ICMSF); A. Koulikovskii and Z. Matyas (WHO). World Health Organization, Geneva, PH/82.37 (Also available in French)

2 *Microbiological Criteria for Foods. Summary of Recommendations of FAO/WHO Expert Consultations and Working Groups 1975–1981*. 1983. Author, J.H.B. Christian. World Health Organization, Geneva, VPH/83.54

Appendix 5

LABORATORIES COLLABORATING IN ICMSF METHODS STUDIES

The following laboratories have participated in one or more projects of the ICMSF testing programs.

Agricultural Research Council, Meat Research Institute, Langford, England
Bundesgesundheitsamt, Postfach 330013, D-1000, Berlin 33, Germany
Canada Packers Research Centre, 2211 St. Clair Ave W., Toronto, Ontario Canada, M6N 1K4
Caravelle Foods, Mississauga, Ontario, Canada
Cattedra di Microbiologia Lattiero-Casearia, Facolta di Agraria, Universita di Perugia, 06100, Italy
Central Station for Dairy Research and Technology of Animal Products, CNRZ, Jouy-en-Josas, France
Centro Latinamericano de Enserenza e Investigacion, de Bacteriologia Alimentaria, Apartado 5653, Lima, Peru
Centro Panamericano de Zoonosis, C.C. 3092, 1000 Buenos Aires
Centro per lo Studio Technologico, Bromatologico e Microbiologico del Latte, Universita degli Studi di Milano, via Celeoria 2, 20133 Milano, Italy
CITIL-INTI, C.C. 157 – (1650) San Martin, Pcia, de Buenos Aires, Argentina
CSIRO, Division of Food Research, P.O. Box 52, North Ryde, NSW 2113, Australia
Danish Meat Products Laboratory, The Royal Veterinary and Agricultural College, Copenhagen, Denmark
Del Monte Corporation, Walnut Creek, California, USA
Department of Animal Science, College of Agriculture, Texas A and M University, College Station, Texas 77843, USA
Department of Food Microbiology, Institute of Nutrition, Budapest IX, Gyali-ut 3/a, Hungary
Department of Food Science, University of Georgia College of Agriculture, Experiment Stations, Georgia Station Experiment, Georgia 30212, USA
Department of Food Science and Human Nutrition, College of Human Resource, University of Delaware, Newark, Delaware, USA
Department of Microbiology, Provincial Laboratory of Hygiene and Prophylaxis, via Mossotti 4, 281000, Italy

Director of Technical Resources, General Foods Ltd., Box 4019, Terminal A, Toronto, Ontario, Canada

Dominion Stores, Toronto, Ontario, Canada

Eastern Region, Food and Drug Directorate, Ralston Building, 1557 Hollis Street, Halifax, Nova Scotia, Canada

Environmental and Public Health Laboratory of Kobe City, 1-5 Kano-cho, Ikutaku, Kobe 650, Japan

Equity Meat Corp., Folcroft, Pennsylvania, USA

Facultad de Veterinaria, Instituto de Industria Animal, Avda. A. Lasplaces 1550, Montevideo, Uruguay

Faculty of Veterinary Medicine, Cairo University, Giza, Egypt

Food Hygiene Laboratory, Central Public Health Laboratory, London, England

General Foods Ltd., Box 4019, Terminal A, Toronto, Ontario, Canada

Health Protection Branch, Health and Welfare Canada, Tunney's Pasture, Ottawa, Ontario, Canada, K1A 0L2

Health Protection Branch, Atlantic Region, Ralston Building, 1557 Hollis Street, Halifax, Nova Scotia, Canada, B3J 1V5

Health Protection Branch, Central Region, 310 Federal Building, 268 Main Street, Winnipeg, Manitoba, Canada, R3C 1B2

Health Protection Branch, Ontario Region, 2301 Midland Avenue, Scarborough, Ontario, Canada, M1P 4R7

Health Protection Branch, Quebec Region, 1001 St. Laurent W., Longueuil, P.Q., Canada, J4K 1C7

Health Protection Branch, Western Region, 1001 West Pender Street, Vancouver, British Columbia, Canada, V6E 2M7

Hokkaido Institute of Public Health, Section of Epidemiology, Sapporo, Hokkaido, Japan

Institut für Hygiene der Bundesanstalt für Milchtorschung, Hermann-Weigmann Str. 1-27, 23 Kiel, Germany

Institut für Lebensmittelhygien, Winterhurerstrasse 270, CH-8057 Zurich, Switzerland

Institut für Milchhygiene und Milchtechnologie, Veterinarmedizinische Universität, Linke Bahngasse 11, A-1030 Wien, Austria

Institut National de la Recherche Agronomique, Laboratoire de Recherches Fromagères, Chemin St. Etienne, 15000 Aurillac, France

Institut Pasteur Du Maroc, Place Charles Nicolle, Casablanca, Morocco

Institut Pasteur Hellenique, 127 Ave. Vassilissis Sofias, Athens, Greece

Institute for Public Health and Medical Research, Pasteur No. 6, Cluj, Rumania

Institute of Food Hygiene, Free University of Berlin, 1 Berlin 33, Bitterstrasse 8-12, West Germany

Institute of Hygiene, Charles' University of Prague, Srobarova 48, Praha 10, Czechoslovakia

Institute Pasteur de Lille, Lille, France

Instituto de Technologia de Alimentas, Campinas, SP-Brasil, CEP 13, 100. Av. Braşil 2.800, Brazil

Jefa de Division Bacteriologia, Sanitaria, Avda Velez Sarsfield 563, Buenos Aires, Argentina

J.M. Schneider, Inc., Kitchener, Ontario, Canada

Laboratoire de Recherches de Technologie Laitière, Ministère de l'Agriculture, INRA, 65 Rue de Saint-Brieuc, 35042 Rennes, France
Laboratorio Provinciale di Igiene e Profilassi, Viale Roma 7D, 28100 Novara, Italy
Laboratory for Zoonoses, National Institute of Public Health, Utrecht, The Netherlands
Laboratory for Zoonoses and Food Microbiology, Rijks Institut voor de Volksgezondheid, Antonie van Leeuwenhoeklaan 9, Postbus 1, Bilthoven, The Netherlands
Landessanstalt für Veterinarmedizin und Lebensmittelhygiene, Berlin, Wilskistrasse 50, 1000 Berlin 37, Germany
Leatherhead Food Research Association, Randalls Road, Leatherhead, Surrey KT22 7RY, England
Microbiology Section, Central Control Laboratory, Nestlé Products Technical Assistance Co. Ltd., Case Postale 88, CH 1814 LaTour de Peilz, Switzerland
Ministerie van Landbouw, Rijkszuivelstation, Brusselsesteenweg 370, 9230 Melle, Belgium
Minneapolis Center for Microbiological Investigation, Food and Drug Administration, Minneapolis, Minnesota 55401, USA
National Bacteriological Laboratory, Stockholm, Sweden
National Centre for *Salmonella*, Central Institute for Scientific Research on Epidemiology, Moscow, USSR
National Health Institute, Department of Health, 52–62 Riddiford Street, P.O. Box 7126, Wellington South, New Zealand
National Health Laboratory, P.O. Box 287, Khartoum, Sudan
National Institute of Public Health, Bilthoven, The Netherlands
Public Health Laboratory, Hull Royal Infirmary, Anlaby Road, Hull, HU3 2JZ, England
Public Health Laboratory, Ministry of Health, P.O. Box 9526, Haifa, State of Israel
Public Health Laboratory of Kanagawa Pref., 52-2 Nakao-cho, Asahi-ku, Yokohama 241, Japan
Reference Laboratory for Environmental Hygiene, Centre for Applied Microbiology and Research, Porton Down, Salisbury, Wiltshire SP4 OJG, England
Research Laboratories, California Packing Corporation, 215 Fremont St., San Francisco, California, USA
St. Ivel Technical Centre, Bradford-on-Avon, Wilts, USA
School of Public Health and Tropical Medicine, University of Sydney, New South Wales 2006, Australia
Shizuoka Public Health Laboratory, 3-6-2 Takajo-cho, Shizuoka 420, Japan
Silliker Laboratories, 1139 E. Dominguez St., Suite I, Carson, California, USA 90746
Silliker Laboratories, Inc., 1304 Halsted St., Chicago Heights, Illinois 60411, USA
Spelderholt Institut for Poultry Research, 7361 D A Beekbergen, The Netherlands
Stichting Centraal Orgaan Zuivelcontrole, Postbus 250, 3830AG, Leusden, The Netherlands
Swift and Company, 1919 Swift Drive, Oak Brook, Illinois, 60521, USA
The Agricultural Institute, Moorepark Research Centre, Fermoy, County Cork, Ireland

The Institute of Nutrition, A.M.S., Solynka 2/14, Moscow, USSR
The Meat Industry Research Institute of New Zealand, Hamilton, New Zealand
The National Food Administration, Box 622, S-751, 26 Uppsala, Sweden
The Pillsbury Company, Research and Development Laboratories, Minneapolis, Minnesota 55414, USA
Tokyo Metropolitan Institute of Public Health, 3-24-1, Hyakunin-Cho, Shinjukuku, Tokyo 160, Japan
Unigate Technical Centre, Abbey House, Church Street, Bradford-on-Avon, Wiltshire BA15 1DH, United Kingdom
Unilever Research Laboratory, Colworth House, Sharnbrook, Bedford, MK44 1LQ, England
Universidad Autonoma de Santo Domingo, Primada de America, Santo Domingo, Dominican Republic
University of Veterinary Sciences, 1400 Budapest, P.O. Box 2, Hungary
U.S. Department of Agriculture, Agriculture Research Center, Food Safety and Quality Service, Beltsville, Maryland, USA 20705
U.S. Department of Agriculture, Consumer and Marketing Service, Beltsville, Maryland, USA, 20705
U.S. Department of Health and Human Services, Public Health Service, Food and Drug Administration, Washington, DC
U.S. Department of Health and Human Services, Public Health Service, Food and Drug Administration, Minneapolis, Minnesota, USA 55401
U.S. Department of Health and Human Services, Public Health Service, Food and Drug Administration, Dallas, Texas, USA 75204
U.S. Department of Health and Human Services, Public Health Service, Food and Drug Administration, Seattle, Washington, USA 98174
West Central Region, Food and Drug Directorate, Federal Building, Main and Water Streets, Winnipeg, Manitoba, Canada
Western Region, Food and Drug Directorate, 1001 West Pender Street, Vancouver, British Columbia, Canada
Zuivelcontrole-Instituut, Postbus 250, 3830 AG, Leusden, The Netherlands

Appendix 6

GENERAL PRINCIPLES FOR THE ESTABLISHMENT AND APPLICATION OF MICROBIOLOGICAL CRITERIA FOR FOODS*

These General Principles are intended to guide, primarily, Codex Committees in the establishment and application of microbiological criteria and to this end they contain definitions of mandatory and advisory criteria which relate specifically to the requirements of the Codex Alimentarius. They are also intended for application where microbiological criteria for foods are being developed.

1 DEFINITION OF MICROBIOLOGICAL CRITERIA FOR FOODS

A microbiological criterion, as defined for Codex purposes, consists of:

1.1 a statement of the microorganisms and parasites of concern and/or their toxins. For this purpose, microorganisms include bacteria, viruses, yeasts and moulds;

1.2 the analytical methods for their detection and quantification;

1.3 a plan defining the number of field samples to be taken, the size of the sample unit and where and, if appropriate, when the samples are to be taken;

1.4 microbiological limits considered appropriate to the food; and

1.5 the number of sample units that should conform to these limits.

2 APPLICATION OF MICROBIOLOGICAL CRITERIA

Microbiological criteria, as defined for Codex purposes, fall into two main categories (see also Section 5 for interpretation):

2.1 *Mandatory criterion*

2.1.1 *A microbiological standard* is a criterion contained in a Codex Alimentarius Standard. Wherever possible it should contain limits only for pathogenic microorganisms of public health significance in the food concerned. Limits for non-pathogenic microorganisms may be necessary and when these are included the provisions of paragraph 6.1 shall apply. A microbiological standard shall

* *Codex Alimentarius Commission, Fourteenth Session, 1981: Report of the 17th Session of the Codex Committee on Food Hygiene.* ALINORM 81/13, Appendix II.

not be introduced *de novo* but shall be derived from microbiological end-product specifications which have accompanied Codes of Practice through the Codex Procedure and which have been extensively applied to the food.

2.2 *Advisory criterion*

An advisory criterion is one of two types contained in Codes of Practice.

2.2.1 *A microbiological end-product specification* is intended to increase assurance that the provisions of hygienic significance in the Code have been met. It may include microorganisms which are not of direct public health significance.

2.2.2 *A microbiological guideline* is applied at the establishment at a specified point during or after processing to monitor hygiene. It is intended to guide the manufacturer and is not intended for official control purposes. It may include microorganisms other than those regarded in 2.1.1 and 2.2.1.

3 PURPOSE OF MICROBIOLOGICAL CRITERIA FOR FOODS

3.1 The purpose of microbiological criteria for foods is to protect the health of the consumer by providing safe, sound and wholesome products and to meet the requirements of fair practices in trade.

4 GENERAL CONSIDERATIONS CONCERNING PRINCIPLES FOR ESTABLISHING AND APPLYING CRITERIA

4.1 The basis of control of microbiologically sensitive foods should be through the application of Codes of Practice. A microbiological criterion should be established and applied only where there is a definite need for it and where it can be shown to be effective and practical. Such need is demonstrated by epidemiological evidence that the particular food is a public health hazard, or where an assurance is required that the provisions of hygienic significance in the Code have been adhered to. The criterion should be technically attainable by good manufacturing practice so that it does not encourage the use of objectionable treatments in an attempt to reduce microorganisms to the acceptable level.

4.2 To fulfil the purposes of microbiological criteria, consideration should be given to

- the evidence of hazards to health;
- the microbiology of the raw material;
- the effect of processing on the microbiology of the food;
- the likelihood and consequences of microbial contamination and/or growth during subsequent handling and storage;
- the category of consumers at risk; and
- the cost/benefit ratio associated with the application of the criterion.

4.3 The number of samples tested shall be as stated in the sampling plan and shall not be exceeded.

4.4 To make the best use of limited resources of money and manpower, it is essential that only appropriate tests be applied to those foods and at those points during the processing and distribution of food that offer maximum benefit in providing the consumer with a safe, sound and wholesome food.

4.5 The need for inspection of the establishment including the production process should be considered.

5 INTERPRETATION OF RESULTS

5.1 When a product fails to meet a criterion the action to be taken depends on the type of criterion and on the circumstances. If the limit exceeded is part of a standard the product concerned must be rejected as unfit for its intended use; if it is part of an end-product specification appropriate action should be taken to rectify the causative factor. It is optional whether any further action is taken. When a limit in a guideline is exceeded this should not necessarily result in rejection of the product but should in general lead to the identification and correction of causative factors.

5.2 When the product is rejected there are in principle several options as to the action to be taken, depending on the findings and the circumstances. Such options include sorting, reprocessing (e.g. by heating), and destruction, and may need to be specified in the criterion. In deciding on the option the major consideration should be to keep to a minimum the risk that unacceptable food reaches the consumer. However, food must not be needlessly destroyed nor declared unfit for human consumption.

6 COMPONENTS OF A MICROBIOLOGICAL CRITERION

6.1 *Microorganisms of importance in a particular food*

6.1.1 The microorganisms included in a criterion should be widely accepted as relevant – as pathogens, as indicator organisms or as spoilage organisms – to the particular food and technology. Organisms whose significance in food is in doubt should not be included in a criterion.

6.1.2 The mere finding, with a presence-absence test, of certain organisms which have caused foodborne illness (e.g. *Staphylococcus aureus, Clostridium perfringens* and *Vibrio parahaemolyticus*) does not necessarily indicate a hazard.

6.1.3 When choosing a test for an indicator organism there should be a clear understanding as to whether the test for this organism is used to indicate an unsatisfactory manufacturing practice or whether it is used to indicate the possible presence of a pathogen. Where pathogens can be detected directly, a test for these should be used instead of tests for indicator organisms.

6.2 *Microbiological methods*

6.2.1 For use in a standard or end-product specification, methods elaborated by international organizations for a food or a group of foods should be preferred. For standards, and wherever possible for end-product specifications, only methods for which the reliability (accuracy, reproducibility, inter- and intra-laboratory variation) has been statistically established in comparative or collaborative studies in several laboratories should be used. While reference methods to be used in standards and end-product specifications should be the most sensitive and reproducible for the purpose, methods to be used in guidelines might often sacrifice to some degree sensitivity and reproducibility in the interests of speed and simplicity. They should, however, have been proved to give a sufficiently reliable estimate of the information needed.

6.2.2 When choosing a microbiological method as a reference method, consideration should be given to the universal availability of media, equipment, etc.

6.2.3 Methods which are applicable uniformly to various groups of commodities should be given preference over methods which apply only to individual commodities. Methods for testing rapidly perishable foods should be so designed that the results of microbiological examinations can be available before the foods are consumed or exceed their shelf-life.

6.3 *Microbiological limits*

6.3.1 Limits should be based on microbiological data appropriate to the food and to the kind of criterion in question. Limits for standards and end-product specifications should be based on data gathered at various stages of production, storage and distribution, while limits for guidelines could be based on data obtained from microbiological monitoring during production. The numerical limits should take into consideration the risk associated with the organisms likely to affect the acceptability of the food, and the conditions under which the food is expected to be handled and consumed. Numerical limits should also take account of the distribution of microorganisms in the food and the inherent variability of the analytical procedure.

6.3.2 If a criterion requires a particular microorganism not to be detected, the size of sample shall be indicated. It should be borne in mind that no feasible sampling plan can ensure complete absence of a particular organism.

6.3.3 Microbiological limits can be related only to the time and place of sampling and not to the presumed number of microorganisms at an earlier or a later stage. As good manufacturing practice aims at producing foods with microbiological characteristics significantly better than those required by public health considerations, a numerical limit in a guideline may be more stringent than in a standard or an end-product specification.

6.4 *Sampling plans*

6.4.1 A sampling plan is the particular choice of sampling procedure and the decision criteria to be applied to a lot, based on examination of a prescribed number of sample units by defined methods. Sampling plans should be administratively and economically feasible. In particular, sampling plans should take into account the heterogeneity of distribution of microorganisms. For standards and end-product specifications, 2- or 3-class attribute plans may find useful applications. (See ICMSF, *Microorganisms in Foods, 2. Sampling for Microbiological Analysis. Principles and Specific Applications*, reprinted 1978.)

6.4.2 Wherever possible, the confidence limits of the sampling plans should be given.

7 SAMPLING METHODS AND HANDLING OF SAMPLES

7.1 The sampling method shall be defined in the sampling plan. The time between field sampling and analysis should be as short as possible and during transport to the laboratory and conditions (e.g. temperature) should be appropriate to the food, so that the results reflect – within the limitations given by the sampling plan – the microbiological conditions of the lot presented for inspection.

8 REPORTING

8.1 The test report shall give the information needed for complete identification of the sample, the results, and the test method.

9 PROVISIONS FOR RECONSIDERATION AT REGULAR INTERVALS

9.1 Criteria should be reviewed and if necessary revised at three year intervals after their adoption by the Codex Alimentarius Commission.

Glossary

acceptance sampling The application of a predetermined sampling plan to decide whether or not the lot meets defined criteria of acceptance.

aerobic plate count (APC) The number of colony-forming units of aerobic mesophilic microorganisms present per gram or per ml in the analytical unit as determined by a standard method.

aliquot The portion of food that is inoculated into a container of bacteriological medium in accordance with a specified method. For example, from a 1 kg sample unit, a 25 g analytical unit may be blended with 225 ml of diluent to give a 1:10 dilution. A 1 ml portion of this suspension would be an aliquot of 0.1 g.

analytical unit That amount of product withdrawn from a sample unit for an analysis.

attributes plan A sampling plan in which each selected sample unit is classified according to the quality characteristics of the product and in which there are only two or three grades of quality: for example, acceptable, defective; absent, present; acceptable, marginally acceptable, defective; low count, medium count, high count.

c The maximum allowable number of defective sample units (2-class plan) or marginally acceptable sample units (3-class plan). When more than this number are found in the sample, the lot is rejected.

case A set of circumstances related to the hazard a food represents and to its anticipated subsequent treatment.

consignment A quantity, large or small, of food destined, in commerce, to a particular recipient, usually consisting of multiple containers of food from one or more lots.

consumer's risk The probability that a lot will be accepted on the basis of results of testing analytical samples when in reality the lot as a whole is substandard relative to the stated criteria of acceptability.

defective Not suitable for the intended purpose.

frame That portion of the lot from which the sample units are drawn. Ideally it should be the whole lot; in practice it may be the accessible portion of the lot.

Good Commercial Practice (GCP) A general term defining conditions of Good Manufacturing Practice combined with acceptable conditions of distribution and storage.

Good Manufacturing Practice (GMP) Those procedures in a food-processing plant which consistently yield products of acceptable quality. GMP is usually described in a code defining processes, equipment, plant layout, sanitation, hygiene, and laboratory tests; but, as used here, it does not apply to any particular code now in use.

hazard The potential to cause (i) disease when eaten or (ii) organoleptic spoilage in a period shorter than the normal shelf-life of the food.

Hazard Analysis Critical Control Point (HACCP) A system in which observations and/or tests are made to identify actual or potential hazards in operations and to identify critical control points in a process. Control measures are designed and implemented, and the control points are monitored to ensure that control is maintained.

indicator Historically, an organism itself non-pathogenic, but often associated with pathogens, used to portray a risk of the presence of pathogens for which feasible methods of detection were not generally available (sometimes called 'index organisms'). This usage is currently expanded to denote groups or species of organisms whose presence in a food reveals exposure to conditions that might introduce hazardous organisms and/or allow their growth. Specific indicators are now used to reveal excessively contaminated raw materials; unsanitary manufacturing practice; contamination from faecal, nasopharyngeal, or suppurative sources; unsuitable time-temperature conditions of storage; or failure of a process. The indicator groups and their methods of enumeration are described in ICMSF, *Microorganisms in Foods, 1. Their Significance and Methods of Enumeration* (1978).

lot In the commercial sense, a lot is a quantity of food supposedly produced under identical conditions, all packages of which would normally bear a lot number that identifies the production during a particular time interval, and usually from a particular 'line,' retort, or other critical processing unit (see Chapter 3, Section A). Statistically, a lot is considered as a collection of units of a product from which a sample is to be drawn to determine acceptability of the lot.

m A microbiological limit which, in a 2-class plan, separates good quality from defective quality or, in a 3-class plan, separates good quality from marginally acceptable quality. In general, values equal to *m*, or below, represent an acceptable product and values above it are either marginally acceptable or unacceptable.

M A microbiological limit which, in a 3-class plan, separates marginally acceptable quality from defective quality. Values above *M* are unacceptable.

microbiological criterion A criterion concerning the presence or quantity of microorganisms or their toxins per unit of mass, volume, or area, established by use of defined procedures and applied in acceptance sampling of foods (see Appendix 6: General Principles for the Establishment and Application of Microbiological Criteria for Foods).

microbiological guidelines A microbiological criterion used by a manufacturer or regulatory agency to monitor a food, ingredient, process, or system; often used also to describe a microbiological criterion where no standard has been prescribed. The Codex Alimentarius definition is more restrictive (see Appendix 6: General Principles for the Establishment and Application of Microbiological Criteria for Foods).

microbiological purchasing specification The microbiological criterion or criteria conditional to acceptance of a specific food or food ingredient by a food manufacturer or other private or public purchasing agency.

microbiological standard A microbiological criterion in a law or regulation controlling foods produced, processed, or stored in the area of jurisdiction of a regulatory agency, or imported into the area.

moderate hazard The hazard associated with ingestion of a food containing a pathogen or toxin which can cause a short-term disease not critically severe in its manifestations and normally without sequelae: for example, salmonellosis, staphylococcal food-poisoning.

Most Probable Number (MPN) A statistically estimated number (per unit of mass or volume) of the test organism present in a sample unit, based upon its presence or absence in replicate aliquots (see also ICMSF, *Microorganisms in Foods, 1. Their Significance and Methods of Enumeration* [1978]).

n The number of sample units which are examined from a lot to satisfy the requirements of a particular sampling plan.

population (epidemiological) The number of people estimated to be at risk from a particular food.

population (microbiological) The number of microorganisms, total or of a species or group(s), dispersed within a defined quantity of food and detectable by the methods used. Hence, a total microbial population or a population of staphylococci, coliforms, etc.

population (statistical sampling) The aggregate of units about which an inference is to be made from analytical results. In this book, population usually relates to the totality of hypothetical individual portions within a lot, each identical in quantity with the defined sample unit.

probability of acceptance The likelihood that a lot of known quality would be accepted on the basis of results of testing analytical samples relative to the stated criteria.

producer's risk The probability that a lot of acceptable quality will be rejected on the basis of results of testing analytical samples relative to the stated criteria.

random sample of size *n* A collection of *n* sample units obtained in such a way that every sample of size *n* has an equal chance of selection, hence excluding bias. Reference to a table of random numbers is usually involved (see Chapter 3, Section C).

sample The total number (one or more) of individual sample units drawn from a lot (ideally at random) which will be tested in accordance with a specific sampling plan and method(s).

sample unit The individual portion or container of food taken as part of the sample and from which one or more analytical units may be drawn for analysis.

sampling plan Used synonymously in this text with *lot acceptance sampling plan*, and meaning a statement of the criteria of acceptance to be applied to a lot, based on examination of a required number of sample units by defined analytical methods.

severe hazard The hazard associated with the presence of a pathogen or toxin in a food which when ingested is likely to cause severe disease, either in a healthy population or in a particularly susceptible group to which the food in question is often destined. Severity of hazard is largely determined by clinical severity of the disease induced, but many ancillary factors are pertinent (see Chapter 5 and Tables 6 and 7). Pathogens of concern in this category include *Brucella* spp.,

Clostridium botulinum, Cl. perfringens type A, *Salmonella typhi, Shigella dysenteriae, Vibrio cholerae* O1 (Table 7); they are normally sought by application of investigational sampling (see Chapter 5, Section N).

shelf-life The period after processing during which a food remains in acceptable condition under appropriate conditions of storage.

stratification A technique for dividing consignments into logical groupings such as size of cans, grades, box, cases, which controls known sources of variation. It may be used where prior knowledge exists that the consignment is potentially not of uniform quality.

stringency The probability that a lot of food of defined microbiological quality will be rejected by a prescribed sampling plan, relative to other sampling plans.

subsample The several individual swabs or portions of food taken from a single carcass or carton and which make up a single sample unit.

variables plan A sampling plan in which the criteria employ statistics of a sample, such as the mean and standard deviation, rather than the segregation of sample units into attribute classes (*see* attributes plans). In the context of this book the variables plans referred to are based on the application of the mean ($\bar{x}$) and standard deviation (s) of the sample log counts. These parameters are combined with a constant or constants (k_1, k_2) selected for the appropriate number of sample units (n) for comparison with a limit value (v or V).

Index